PASS THIS BOOK ON

Once you have read this book I would like you to pass it on to a friend, neighbour or relative. Think of someone who might find it useful. Write your name and the date you passed it on here. After they have done with it, ask them to sign it here too and then pass it on to someone else they think will benefit. And so on.

Let's look at some numbers.

Let's say this book sells 1,000 copies. If those 1,000 people who bought it then give up buying 50 bottles of water in a year, instead refilling a water bottle from the tap, and all pass on their copy of the book to another 1,000 people who then pass it on to another 1,000 people – all of whom give up using 50 plastic bottles a year – we've already saved 150,000 bottles from going to landfill, recycling or the ocean.

In 10 years that would be 1,500,000 plastic bottles not bought and thrown away.

Who says individual actions don't matter?

My name is: ....

I gave this book to: ....

Date: ....

My name is: ....

I gave this book to: ....

Date: ....

My name is: ....

I gave this book to: ....

Date: ....

My name is: ....

I gave this book to: ....

Date: ....

My name is: ....

I gave this book to: ....

Date: ....

My name is: ....

I gave this book to: ....

Date: ....

NO. MORE. RUBBISH. EXCUSES.

How to reduce your waste and why you must do it now

MARTIN DOREY
Founder of the #2minutebeachclean

EBURY PRESS

Martin Dorey is a writer, surfer and beach lover. He founded the Beach Clean Network with Tab Parry in 2009 and started the #2minutebeachclean hashtag in 2013 after North Atlantic storms left UK beaches littered with plastic rubbish. It's a simple, effective idea: pick up litter for two minutes, bag it, tag it, bin it and recycle what you can. In 2019, the Beach Clean Network became The 2 Minute Foundation, a charity devoted to cleaning up the planet two minutes at a time. The #2minutebeachclean campaign has now been extended to include the #2minutelitterpick and #2minutestreetclean with more than 800 beach cleaning and litter-picking stations around the UK and Ireland. In Martin's first book, *No. More. Plastic.*, he continued the clean up by introducing the idea of the #2minutesolution to encourage people everywhere to make those vital, small changes that can add up to make a big difference.

CONTENTS

HELLO.

Thanks for opening this book. I hope you enjoy reading it and I hope you enjoy implementing changes, because that's what we need right now.

Don't get me wrong: changes have been made, and they will continue to be made in the years to come. But if there's one thing I have found since I started campaigning about plastic, it is that changes have to come from us.

TV programmes like *Blue Planet II* may open our eyes to the vast problem of plastic pollution in our oceans and the way our society allows it, but it is up to us to make the vital changes we want to see in the world. Why? Because we are the ones with the money to buy into that which we think is right.

It's as simple as that.

However, the post *Blue Planet II* world is a confusing, cluttered place and it's not easy to navigate it.

Hence this book.

I have written it because I feel it too. I am a surfer, ocean lover, writer, coastal dweller and green activist. And even I have a problem with getting rid of plastic and dealing with the waste in my life. I find choices tougher than ever and I find dead ends and red herrings wherever I turn.

I first became aware of the problem with litter in the ocean in the 1980s when I surfed in North Wales while at university in Manchester. There wasn't as much plastic back then, just a lot of aerosol cans that washed up and then exploded on our beach fires. When I moved to Devon in 1996, I noticed hundreds of bottles of Hawaiian Tropic sun lotion on every beach I visited. I later found out that those bottles came from a cargo spill. That was the first time I became fully aware of the issue of ocean plastics.

In the ensuing years I attended beach cleans (events where a community comes together to remove plastic and litter from beaches) and then, in 2007, I visited the beach below Sloo Woods in North Devon where I found a natural collecting point for ocean plastic. The beach, a narrow and steep strip of cobble, was knee-deep in plastic bottles, crates and lost fishing gear. Angry and heartbroken I vowed there and then to do something to make a difference. I organised a beach clean with my children's school and with help from my friends at the National Trust and Northam Burrows Ranger Service. Then, later that year, with some web developers and my friend Tab Parry, I set up the Beach Clean Network, an online service to unite beach clean organisers with volunteers.

We bumbled along, attending beach cleans and organising them too. I moved to Bude. Then, in the winter of 2013, ferocious Atlantic storms brought havoc to our coastline. Beach huts were wrecked and the detritus of many years of out-of-control plastic production and careless use spewed forth onto our beaches from the depths. I took a bag to fill with rubbish and then went back with another, and another. My fellow beach lovers joined in too. I took pictures and posted them to Instagram using a new hashtag, #2minutebeachclean, hoping that someone might see them and do the same. They did.

The idea of taking two minutes out of your day to improve your local environment seemed to resonate. I founded the #2minutebeachclean hashtag and, together with the gentle and guiding hand of Dolly and a growing team of devoted activists, built a movement around taking two minutes out of your day to improve your immediate environment. The message – whether you use the #2minutebeachclean, #2minutelitterpick, #2minutestreetclean, #2minutesuperhero or #2minutesolution hashtags – is always positive. You might feel daunted by the scale of the problem, but you can make a difference with your small, everyday actions.

Six years later, the #2minutebeachclean has become a global movement, with over 123,000 posts to Instagram using the hashtag. We also have a network of over 800 #2minutebeachclean, #2minutelitterpick and #2inutestreetclean stations around the UK and Ireland from which the public can borrow litter pickers and bags to help clear up. In trials on one beach in Bude, the presence of the station resulted in 61 per cent less litter picked up on monthly beach cleans compared with the year before.

The #2minutebeachclean campaign was top of a list of 'things you can do for ocean conservation' on the *Blue Planet II* website in 2017, something my team of eight dedicated activists is eternally grateful for. It has brought us to the attention of big business, local councils and people all across the world.

As a result of the campaign, I have lectured in Norway, Holland, at Stormont in Northern Ireland, in Dublin, London and at a TedX event in Kiel, Germany, as well as at conferences, book festivals, schools and universities. My bestselling first book on plastic, *No More Plastic,* was published in 2018. My second, a book for children called *Kids Fight Plastic,* was published in 2019.

As a campaigner, I have been on the front line of the marine litter crisis. I have found birds that have been killed by lost fishing tackle, attended the autopsy of a dolphin that was butchered as bycatch (unintended and untargeted catch in a fishing net), visited a remote gannet colony where the nests were made from nylon fishing net and seen the scars left around a seal's neck from when it was strangled by a nylon rope.

Plastic has now been found at the bottom of the Mariana Trench, the deepest part of the ocean. It's been accumulating on uninhabited islands thousands of miles from civilisation. It's been entangling and killing animals in their hundreds of thousands. It's in shellfish, bottled water and every river in Britain. It's pouring out of estuaries in Asia. It's in our homes, our cars, our shops and in every aspect of our lives. It's in our tea bags and tampons. Our tea bags and tampons! Think about that. The chemicals from plastics are in our newly born. Babies crawling on carpets weaved from man-made fibres may be breathing in more plastic than anyone else. It is everywhere.

When *No More Plastic* was published it was a direct response to the plastic crisis, as we saw it then. But now it seems clear that it isn't just plastic that's the problem; it's the wasteful way in which we live. When 10 million pumpkins are grown to carve into scary lanterns at Halloween in the UK each year, and yet 60 per cent of them will not get eaten, it's time to get some perspective.

But, hey, it's only one pumpkin, right?

The more you follow the plastic debate, the more it leads you down other roads: to waste, food production, climate change, industry, the exploitation of our lovely planet for financial gain and the way in which we live.

So I feel it's time to update the story. It's time we took it to the next level and begin seeing plastic as just a small part of the waste crisis. Along the way we can try to clear up some of the confusion about new products and practices that have come about because of the crisis, and have caught the public's attention because of it, yet may simply be adding to the problem.

A lot has happened in the short time since *Blue Planet II*. I still believe there are things you can do, on a personal level, every day, that will make a difference. In the pages of this book I will show you ways you can cut even more plastic from your life. I will also try to point you in the right direction when it comes to making difficult choices as some solutions may turn out to be too good to be true.

So it's time to sharpen the resolve, question everything and keep on keeping on.

Everything you do makes a difference.

HOW TO USE THIS BOOK

In the following chapters, you will find a series of ideas for changes and swaps you can make to reduce what you throw away each year. They all add up to make a big difference. Each and every action you take to reduce your waste, reduce your impact and rethink the way you live will send out ripples, however seemingly insignificant, that will go out into the world and have a positive effect, even if you never see it.

In this book I will help you make better planet-positive decisions when it comes to bathing, shopping, eating, clothing and living. These sections of practical advice and suggestions as to how you can reduce your impact are headed 'Time to Act'. In my first book, *No. More. Plastic.*, each of these little changes was called a #2minutesolution. Similarly, the solutions I suggest rarely take more than a few minutes to begin implementing, and yet add up to make a huge difference.

Let's start here...

- REFILL your water bottle, at the very least.
- REUSE your possessions as much as you can, instead of throwing them away.
- REPAIR, REVIVE and RESTYLE your possessions to love them for longer.
- REDUCE your waste, RESIST accumulating more stuff and RE-EVALUATE and RECONSIDER the way you shop.

We need to RETHINK our relationship with plastic and waste and REBEL against those who seek to push it upon us, so we can REDRESS the balance in nature.

And then, when we've done that, we need to RECYCLE what's left.

OUR POST-BLUE PLANET II WORLD

It's 17 February 2018. The night is still, with a light breeze coming from the north east, bringing a small chop across the bay. The waves slap against the sea wall while a lone mast, swaying gently, clanks in a slow rhythm to the movement of the water. It's dark, with a new moon barely visible behind wispy, high clouds. Clovelly, a tiny village on the north coast of Devon, is quiet. A few lights illuminate the steep, cobbled, car-free streets and the tumble of fishermen's cottages.

A couple of miles away is the beach below Sloo Wood, a stretch of sand and cobble where, more than ten years earlier, I found a cove knee-deep in plastic bottles, nets, broken pieces of fishing crates and plastic drums.

From Sloo Wood, on a night like tonight, you can see nothing but the black shape of the wooded cliffs across the bay and the occasional streetlight in the combes of Bucks Mills and Clovelly. Though if you were to look hard enough you might just about make out the light from a bedroom window.

Tonight, there's only one person awake, Dolly. She's sitting up in bed, under her duvet, her tattooed arms holding her phone in front of her. Occasionally she pulls her wild, brown hair away from her face to better see what's unfolding before her. It's nearly 3am in the UK, which means America has been awake for some time, but they have only just begun to watch the final episode of *Blue Planet II*. Dolly, who has already seen the series, is watching the #blueplanet2 hashtag on Twitter. Fourteen million people watched the first episode of the series when it aired in the UK in late 2017 and around three million are watching now, Stateside. She knows what's coming.

Sure enough, as Sir David Attenborough delivers his final, withering warning to the people of the world about the state of the oceans, America wakes up. After a good start cooing over nesting turtles and marvelling at orcas sharing their catch with the Norwegian fishery, the mood has changed. America is now shocked, despairing at what it has just seen: a whale eating plastic, bleached coral reefs, albatross chicks on remote shores dying with stomachs full of plastic. People are using the programme's hashtag to express their horror at what we've done to our oceans. They admit to using single-use cups, plastic bags, toothpicks and cutlery: exactly the items they have just seen killing that albatross chick. They realise, as one, that it is us. We are responsible.

Now they have awoken, they want to do something to make amends. But they don't know where to start. Dolly, mother of five, grandmother to three, mother hen to an online following of 50,000, plus many thousands more onlookers, is there to guide them to the #2minutebeachclean project. She invites them to join our community and find out what they can do for ocean conservation.

There is no blame or browbeating, just positivity and love. Whatever you can do matters, she tells them, little changes add up to make a big difference. Suddenly, people are ready to listen to our message.

Blue Planet II wasn't just the turning point for the #2minutebeachclean project, it was the wake-up call that we all needed to start doing something, anything, to reverse the awfulness of the state of our oceans. We couldn't stomach the indiscriminate killing of whales, seabirds, seals and dolphins any longer. We realised then that it was up to us to make changes.

We bought reusable coffee cups. We stopped using plastic bags. We signed petitions. We shared memes of kids saving the oceans and made internet superstars of litter picking grandads. We saved a turtle with every bracelet. We did what we thought was right, our good intentions guiding us.

And then what?

The plastic landscape has changed greatly since 2017. Where once just a few lone voices and organisations were campaigning to raise awareness and get the attention of the wider general public, while scientists investigated plastic in the oceans, microplastics and the way plastics interact with wildlife, now it's a cause célèbre. Industry is waking up, at last, and governments, though still overly concerned with business and economics instead of the health of our planet, are making slow, lumbering moves towards consultation and half-arsed solutions. Businesses, we are told, are working on it. Sadly this doesn't necessarily mean doing it.

And yet, despite this great news, plastic production has been on the rise since 2017. According to a report from the Center for Environmental Law: 'In 2019, the production and incineration of plastic will produce more than 850 million metric tons (Mt) of greenhouse gases – equal to the emissions from 189 five-hundred megawatt coal power plants.'

They also claim that, at current rates, by 2050 it will rise to 2,800 million metric tonnes (Mt).

The big oil companies are spending billions on petrochemical plants to flood the world with plastic packaging. And we think we're saving the world by using a reusable coffee cup.

It's time to up our game.

YOU MAKE THINGS HAPPEN

There are all kinds of people in the world, ranging from those who think that it's theirs to exploit for their own benefit, to those who would – and in some cases do – die for their planet earth. You, I suspect, fall somewhere in between. Most of us are a bit like that. We don't want to stand by and do nothing and yet we don't know if we're ready to chain ourselves up in front of the bulldozers.

As long as you are doing something, even if you are at the very beginning of your journey, it's something.

The most important thing is that as many of us as possible do something, anything, because doing something is better than doing nothing. If you do nothing, nothing happens. You need to find your own way of living in a way that puts less pressure on the planet, make a success of it and then move on to the next challenge.

DON'T JUST STOP THERE

Don't be content to give up plastic straws and then carry on as before. That's not the idea. The plan is that you do one thing at a time, making more difficult choices the further you get. You can do it because you've done it before. If you can give up plastic straws you can live without sausages in black plastic trays. And if you can live without them you can learn to make going to your local greengrocer or farmers' market a part of your lifestyle.

You don't have to be perfect. You just have to try.

Hopefully, by the time you finish this book you'll be able to recognise some of the brands, food, clothes and ways of living that are planet kind and it'll become easier to navigate your way to a happier, more planet-positive way of living. And the more you do, the better our world will become.

THEY DON'T MAKE IT EASY FOR US, DO THEY?

Our society doesn't make it easy for you to avoid plastic and live a cleaner, less wasteful life. Capitalism and globalisation make it possible for us to get anything we want at a moment's notice, never mind the cost to your pocket or the planet. Advertising promises us a better life, if only we buy the product they are flogging this week. It also makes us feel inferior and inadequate when we don't buy the latest product, services or go to the best holiday destinations.

On top of this, social media makes us feel as if everyone else is having a better life, having more and having it more often (they aren't). Mobile phones take amazing pictures that we can share in seconds. The phenomenon of FOMO (fear of missing out) hits hard when you're stuck inside studying for your GCSEs or having to work a zero hours contract in a call centre somewhere.

It's no wonder that 1 in 8 of 5–19-year-olds are dealing with a mental health disorder of some kind.

When it comes to fashion, cars, phones, TVs, gadgets and gizmos we are led to believe that the latest is the best. In-built obsolescence renders them undesirable or useless when the new model comes out so you have to keep consuming more and more.

It's a vicious cycle, cynical at best, designed to make money, not to make you happy. And the sooner we realise that, the happier we'll be.

RESOLVE IS YOUR GREATEST WEAPON

In the fight against waste and plastics – and anything else that threatens our planet – the greatest weapon you have is your resolve.

Whatever motivates you – cleaner beaches, saving wildlife, keeping your neighbourhood clean, leaving a lasting legacy for your children, reducing the effects of climate change – make sure you use it to keep going. It's the fire in your belly that you cannot extinguish.

Nurture the flame and keep it burning bright to guide you.

Remember: what you are already doing – as long as you are doing something – is good. We all need to do more, and we will, but that isn't going to please everyone. Some hardline greens, and this is not all of them, have been known to look down their noses at other people because, well, 'they aren't doing enough'. I call this greenupmanship. There will always be someone who knows more about green issues, recycles more, is vegan, goes on more marches and isn't afraid to let you know about it. Ignore them.

As long as you are doing something, even if you are at the very beginning of your journey, it's better than nothing. If you are new to the plastics crisis then let's call this moment – now – the start of something wonderful. If you are an old hand, then I hope you will still find something to enlighten you and take your planet-positive journey just that little bit further. If you find you do it all already, fantastic, you are my hero, but please pass this book on to someone who doesn't.

WHERE DOES OUR WASTE GO?

TROUBLE AT THE RECYCLING STATION

In *No. More. Plastic.*, I asked the question: is recycling the answer? What is even clearer now is that, while we must continue to recycle as much as we can in order to support a new, more sustainable system, we cannot simply recycle our way out of the waste problem.

We think of recycling as a safe and responsible place where we can send all our waste in the belief that it will be turned into something useful later. Once it's out on the kerbside it's off our consciences. It's very convenient for us to put it out and carry on as before, having done our bit. With recycling we can continue to consume, buy with impunity and have whatever we want, while also believing we're doing the planet a favour by washing out the odd yogurt pot.

We can't. Recycling is better than landfill, of course, but the process uses energy, costs money and often doesn't give us a like-for-like product at the end. Recycling is downcycling, unless it's glass or aluminium (more on that later), which means that we're getting an inferior product out of our yogurt pot. It can only go so far.

According to a global report on plastics that came out in the journal Science Advances in July 2017, only 9 per cent of all plastics have ever been recycled. And only 10 per cent of that has been recycled more than once. At this rate, it seems extremely unlikely that we will be able to recycle what we make and discard, now or in the near future.

One group of global corporations, the Alliance to End Plastic Waste, has made a pledge of £1.5 billion to stop plastic waste. The money they pledge will support small businesses and incubator projects to develop and support recycling projects (in Asia in particular). While this is vital I think we need to read between the lines here. The list of companies supporting the project reads like a rogue's gallery of plastic producers and includes BASF, Chevron Phillips Chemical Company LLC, ExxonMobil, Formosa Plastics Corporation, Mitsubishi Chemical Holdings, Procter & Gamble, Shell and Total. It makes me think that this isn't about cutting plastic production, it's about cleaning up and placing the responsibility on the end user in order to be able to continue producing plastic with impunity. Some might call it greenwashing.

It suits big businesses to support recycling schemes because it places the onus on consumers to be more responsible, to make more effort to sort out the mess and, ultimately, makes us pay for it through local taxes. Yes, that's right. Your council tax gets spent on taking your recycling away, whether you like it or not and whether you use packaging or not. Really the solution we should be looking at is a drastic reduction in plastic usage and then recycling only what is left, not an increase in production and more recycling to cope with the extra waste.

HOW CHINA'S NATIONAL SWORD CHANGED WORLD RECYCLING

Until December 2017, China's waste recyclers accepted waste (and were paid to take it) from the West, including the UK, USA and Europe. While some went to other Asian countries, the majority of it went to recyclers in China and Hong Kong. This was because of a general rise in recycling in Europe and the USA, and a rise in manufacturing in China from the 1990s to the 2010s – ships bringing goods from China went home filled with highly valued recycling, which made good economic sense and made China the perfect place for recyclers to build empires out of our scrap.

However, these good times were thwarted – firstly by China's 2013 Green Fence policy, an initiative that lasted a year, which imposed stricter conditions on the quality of recycling, its environmental consequences and sought to stop illegal imports and processing. Next came the National Sword policy in 2017, a positive effort to clean up the recycling industry. Recycling imported without a proper licence was seized and certain types of recycling such as mixed paper and low-grade plastics were banned. During the course of 2017, China went from accepting just about anything to only accepting the cleanest, highest-value recyclable waste.

Suddenly, in the UK, our waste had nowhere to go.

The knock-on effect of the 'ban' was that we sent our recycling to other countries in Asia and Europe (as gateways to Asia) because it was the cheapest solution to our waste crisis. Malaysia, Thailand, Indonesia and Vietnam (among others) started to take our recycling, legally and not, where it was processed, sometimes at great risk to the environment because of the quality of the materials and practice.

In 2019, other Asian countries started sending our recycling back to us because they could not deal with it properly. The BBC reported that Malaysian Minister Yeo Bee Yin said, 'What the citizens of the UK believe they send for recycling is actually dumped in our country.'

Our recycling is being exported, sifted for the most valuable plastics and then dumped or burned in someone else's country. It is causing environmental damage and entering the oceans via their rivers.

The recycling we can no longer export is recycled locally, goes to landfill or gets burned with general waste in energy-from-waste incinerators.

This is from the UK government's annual figures for waste and recycling in England:

- Recycling waste from households went up in 2017 to 45.2 per cent (a 0.3 per cent increase from 2016).
- Total waste from households decreased by 1.5 per cent in 2017 to 22.4 million tonnes (this is equivalent to 403kg or 888lbs per person).
- Overall, the amount of waste created in 2017/18 remained about the same compared with the previous year.
- However, while landfill volumes went down and recycling rates stayed about the same, incineration volumes went up.

WHY DON'T WE RECYCLE ALL OUR OWN WASTE?

For a start, recycling is complicated. Some products are made of a combination of different types of materials (known as a composite), making them harder and more expensive to clean, separate and recycle.

The best quality recyclate is that which is pure, clean and high in value, like HDPE milk bottles. According to my local recycler, Coastal Recycling, HDPE milk bottles are a good size to sort, often clean (because we do our duty to wash them out before they go out on the kerbside) and make good raw material for other products.

Some things that could be recycled are too small to be picked up. Things like plastic straws escape the process.

Contamination from food waste is another problem. UK councils often ask for non-contaminated food trays (the advice is that plastic food trays should be rinsed, not washed, and not in a dishwasher) because contamination risks rejection by recyclers. This includes grease left on pizza trays and yogurt residue left in your pot.

The wrong type of plastics (for example bio plastics) getting into a recycling batch can render the entire batch worthless. This means your recycling goes straight to landfill or to energy recovery (this is when rubbish is burned to make energy) where its journey will end.

Because of the range of materials used by manufacturers, the symbols for recycling get confusing. Plus, as we discussed in *No. More. Plastic.*, different local authorities are able to recycle different things. A 2019 report by Recoup, the charity that promotes recycling plastics in the UK, claimed that 35 per cent of respondents questioned in a survey felt unsure of what items could be recycled. The common factors they claimed hindered recycling were:

clear instructions from local authorities on what constituted 'food contamination', how it affected recyclability, recycling versus green dot icons (this is a sign that the product's manufacturers have contributed to a recycling responsibility scheme), having to split packaging into different types of materials (a sleeve on a drinks bottle, for example, which must be separated before recycling).

The UK is set to miss EU recycling targets of 65 per cent for 2035 by a country mile, according to some reports, with barriers to recycling stated as being consumer confusion, an over enthusiasm for burning waste for energy, lack of infrastructure and investment (thanks again austerity) and cost.

It seems clear that, even if we all recycled perfectly, we wouldn't have the capacity to deal with it anyway. The rest of the world doesn't want it and we can't let it infiltrate the environment, sit in landfill or emit greenhouse gases. Besides, even if we recycled all of our waste, it's still degrading, being compromised and contaminated by other materials, meaning we may still have to rely on virgin raw materials, whether that's oil-based plastics or bioplastics. One requires fossil fuels and the other requires land needed for agriculture.

CAN'T WE MAKE NEW STUFF OUT OF RECYCLING?

While moves are in place to develop food-grade recycled bottles from old bottles, for example, the quality of the raw materials means plastic gets down-graded each time it's recycled. This means it's often not possible to recycle like into like. A plastic drinks bottle cannot be recycled into an identical drinks bottle (unlike it's a glass bottle, which can be recycled into another glass bottle).

WRAP (Waste and Resources Action Plan) has been working with the dairy and plastics industries to commit to producing a 50 per cent recycled HDPE milk bottle by 2020. Likewise, the Coca Cola company have said they will launch products in drinks bottles made from 100 per cent recycled PET in Europe and are on course to make bottles across their ranges at least 50 per cent recycled content by 2020. However, let's not get too excited. The other 50 per cent has to be virgin plastic, which still comes from oil production, or from bio plastics which come from 'sustainable' sources (but perhaps taking land away from food production or forests).

Meanwhile, others are claiming, at last, that their bottles or products are 100 per cent recyclable, which is great, but doesn't get away from the fact that the recyclability relies entirely upon the system that's needed to recycle it and the commitment required to recycle it. Again the onus is on us.

CAN WE UPCYCLE OUR WAY OUT OF THIS MESS?

How about upcycling our waste into something else? Is this the solution to the problem of unrecyclable materials? Reading headlines like 'Can plastic roads save the planet?' (BBC, 25 April 2017), you'd be led into thinking it is.

Plastic Roads: this is an idea that's been developed for the greater good, not for profit in India, and in other parts of the world too, but that comes with significant issues. Firstly, plastic emits toxic chemicals and greenhouse gases when heated, which might not be great for those laying it. There are also concerns around the durability of the surface, which contains just 0.5 per cent plastic from 'household waste' (although plastic bottles and bags are unsuitable for use in the mix). With wear and tear, could it become microplastics in years to come?

Eco Bricks: eco bricks were all the rage in 2019. The idea is simple: you take your unrecyclable soft plastics – like crisp packets and plastic wraps from salad – and you stuff them into a plastic bottle, which can then be used to make buildings and other constructions in developing countries. The problem I have with eco bricks is that they don't do anything to halt the flow of plastic or our consumption of it, even though they do sequester plastic and protect it, to some extent, from degradation. However, we mustn't forget that plastic has been proven to give off greenhouse gases as it degrades in the environment. Long-term use would need to address these issues. Better not to use it in the first place.

Furniture, decking and other stuff: while you might describe it as upcycling, creating new things – street furniture for example – out of plastic is downgrading it. It'll have a useful life for a while, but what happens after? Will it degrade, splinter, decay and decompose? No, say the manufacturers. Can it be recycled once it's not needed any more? No one will say.

Time to Act:

Recycle as much as you can: recycling is not the answer, but we still have to recycle as much as possible. Our part in this is to make sure what we put into our recycling bags is right for our local authority and is free from contamination. Get in touch with your local authority to find out what you can and can't recycle locally and what kind of state it needs to be in for best results. Given a choice between sending plastic to landfill or recycling, it has to be recycling.

Reduce your impact: when you cut down on your plastic use you automatically reduce your waste and eliminate the problems associated with waste, transport and the after-effects.

Request more transparency: ask your local council if they can provide you with an audit trail for the recycling you send to them. Have they got a 'clear line of sight' between your products and the end of their life? Do they sell their plastics and waste to reputable recyclers and brokers? Is it recycled here in the UK? Question everything. If there are no clear answers for the products you buy, switch to products with a guaranteed end of life.

Rebel!: recycling should be standardised so everyone everywhere knows what to do with theirs. Labelling should be clear and non-recyclables banned. Let your MP know. Tweet them. Sign petitions.

Respond to manufacturers: demand that manufacturers stop using non-recyclable packaging or packaging that's difficult to recycle. Write to them. Tweet them. Send their products back.

Reject: buy products that have recycled content and that are also recyclable, in order to push up the value of recycling. If it says 'cannot currently be recycled' (or similar) don't buy it. If it's PET or HDPE then there is a chance it will be recycled as a valuable resource. Composites are tougher to deal with. Just because something says it's recyclable, doesn't mean it gets recycled or that infrastructure exists to recycle it.

PLASTIC AND PLASTIC ALTERNATIVES

If you're going to look at waste, the first and most obvious place to start is plastic. It is still a huge problem and will continue to be as plastic production ramps up and alternatives hit the market. And yet . . .

THE PLASTIC FANTASTIC

Plastic is a remarkable man-made substance that has all kinds of wonderful uses. It has changed the world. It has saved lives, made medical procedures possible, given us phones and tablets, PCs and iPads. It's cheap, mouldable, strong and durable. In my office now, an office devoted to cleaning up ocean plastic, two minutes at a time, I cannot avoid being surrounded by screens, pens, cables and printers. I am looking at my plastic screen through plastic glasses and tapping away at a plastic keyboard. It's impossible to avoid.

It is almost impossible to separate plastic and consumerism. The two have been on a collision course with nature ever since the 1950s, at a time when our society was promised an easy life. With disposable plastic products like plates and cutlery being pushed upon us, who could argue? Life was hard before the fifties. The brave new world promised convenience.

Disposability – single-use – became the big thing. Today this has never been truer: disposability appeals to our lazy side, meaning we can eat on the run, discard the waste and carry on without a thought for anything other than our next appointment. Why stop for a coffee when we can get one to take out? And why should a coffee chain employ dishwashers, or even fork out for a dishwashing machine, when they could let someone else clear up (and pay for it)? There's money to be made! And we, in our vanity, carry coffee cups like a lifestyle accessory.

We don't question the use of those hard-to-get-into plastic hanging packs either, even though their sole purpose is merchandising (being able to hang an item on a wall so that customers can see it and find it easily).

We don't question delicious looking fruit nestling in bubble wrap (so it can be transported thousands of miles, at great cost to the environment).

We are so used to having a pizza covered in plastic wrapping.

Because we like our lives to be easy.

Our forefathers would be shocked by the lack of thrift today. My grandfather, who never threw anything away, grew vegetables and cut up leaky rubber gloves into rubber bands, would have been dumbfounded by today's waste. His daughter though, my mother, thinks it's great that we can have it all, right now, when she remembers not being able to have anything. The rest of us have never known anything else.

The make-do-and-mend culture of the war years came to an abrupt end in the 1950s as we entered the 'Golden Age of Consumerism'. Disposability makes for a continuous market. The motor industry had already planned obsolescence (see page 197) in the 1930s, and the nature of plastic, and our appetite for new, luxury items and no hassle clearing up, made it easy for companies to keep selling the same stuff over and over again.

The idea that we can continue to produce and consume more and more stuff indefinitely is absurd. A report in 2017 from the Centre for International Environmental Law claimed that scientists became aware of the problem of plastic waste in the ocean in the 1950s. The report also said that the chemical and petroleum industries were aware of, or should have been aware of, the problems caused by their products by no later than the 1970s.

Is it any surprise that the Keep America Beautiful campaign, which was funded by drinks and packaging producers to shift the onus for waste onto the consumer was formed in 1953?

The plastics industry, instead of working to help, has, as far as I can tell, adopted a siege mentality, opposing plastic bag bans, lobbying against deposit return schemes and rubbishing research into the spread of plastics. If we reuse products, buy stuff to last, or mend their products when they break, how will they continue to make profits?

Interestingly though, in 2019 two of the world's biggest producers of single-use plastic, Coca Cola and Pepsi, distanced themselves from the Plastics Industry Association, the group lobbying for plastics in the USA, because they felt they were no longer aligned. Let's hope this leads to great changes. We'll be watching.

Plastic is a part of our lives. The problem with it is us. It's cheap so we use it instead of more expensive natural materials. It's durable so we use it in places where it will last, in order to reduce maintenance costs, not actually thinking it will outlast us all. It's light so it's cheap to transport. It's mouldable so we use it in toys, bottles, packaging and almost every kind of product. Our society, with its hunger for the new, the cheap, the fresh and the exotic, makes it irresistible and unavoidable and fills our lives with toxic, vapid and, ultimately, often useless stuff.

And yet, we haven't worked out what to do with it all. We haven't thought about what will happen to it when we finish with it. We haven't stopped to consider what its 'end of life' will be. What will we do with it? Where does it go? How can we effectively dispose of it?

We rely on recycling or clever ideas to reuse or remake it, forgetting that it degrades and can't be reprocessed infinitely, that it is persistent and will never go away, that we don't have the infrastructure to cope with it all, that it fractures and fragments, that it releases harmful chemicals and greenhouse gasses and that, unless we deal with it properly and stop using so much of it – and by that I mean deal with the end of its life – it is a disaster waiting to happen. Sadly, it's been this way for a long time, which is why we are where we're at. Plastic toys that came free in cereal boxes in the 1950s still wash up on Cornish beaches from time to time, largely intact.

WHAT'S NEW? THE PLASTIC UPDATE

After the world saw the turtle with the straw up its snout and the seahorse with its tail wrapped around a cotton bud stick, things did happen:

- In the UK, supermarket chain Waitrose declared in their Food and Drink Report for 2018–19 that 'nearly 9 in 10 people (88 per cent) who saw that episode of BBC's *Blue Planet II* about the effect of plastics on our oceans have changed their behaviour since'.

- In 2018, 'single-use' was dubbed word of the year by Collins dictionary.
- Members of WRAP's UK Plastic Pact – which include the world's biggest players but no oil companies – committed to making 100 per cent of plastic packaging reusable, recyclable or compostable by 2025. The Plastic Pact sets out the following aims: that 100 per cent of plastic packaging should be reusable, recyclable or compostable by 2025; that 70 per cent of plastic packaging should be effectively recycled or composted by 2025; that single-use packaging will be eliminated by 2025; that 30 per cent of all plastic packaging will be recycled content by 2025.
- The UK government's environment plan set a target of zero avoidable waste by 2050.
- The BBC announced that they will cut all single-use plastics by 2020.
- In 2018, the UK government banned microbeads in cosmetic products.

- In October 2018 the EU declared it would implement a ban on single-use plastics, possibly as soon as 2021.
- Global Web Index reported that over 50 per cent of consumers said that they've reduced the amount of disposable plastic they use.
- In 2018, Boston Tea Party, a small West Country coffee firm, said that it had stopped giving out takeaway coffee cups because of their environmental impact. It cost them £250k in revenue to take such a bold stance. Bravo.
- High street chains The Body Shop and Holland and Barrett stopped making and selling wet wipes over concerns about their environmental safety.
- Morrisons replaced plastic bags in their fruit and veg aisles with paper in 2018, because, 'We've listened to customers' concerns about using plastic bags for fruit and vegetables and that is why we are bringing back paper bags.'

- In late 2019, Sainsbury's introduced reusable vegetable bags for bagging up loose veg. (And yet continue to sell the majority of their veg wrapped in plastic.)
- Unilever's Sustainable Living Plan sets targets to halve waste and ensure all plastic packaging is fully reusable, recyclable or compostable by 2025 as well as increasing recycled plastic content in packaging to 25 per cent. We wait with baited breath.

AROUND THE GLOBE

In the last few years there have been some bold stances taken against plastic around the globe. According to Statista, 66 per cent of countries around the world now have legislation relating to plastic bags while only 14 per cent of countries have legislation restricting single-use items like straws, cutlery and plates.

2002 BANGLADESH banned plastic bags.

2008 RWANDA and CHINA banned plastic bags.

2015 UK places a levy on plastic bags of 5p.

2016 HAMBURG banned plastic coffee pods.

FRANCE banned plastic bags and cutlery.

MOROCCO banned plastic bags.

2017 KENYA banned plastic bags and introduced a fine for everyone selling, carrying or using a plastic bag.

ZIMBABWE banned all polystyrene food containers.

NEW DELHI banned disposable plastic.

2018 ALBANIA and LITHUANIA banned plastic bags.

GREECE placed a charge on plastic bags given out in shops.

VANUATU banned plastic bags, polystyrene takeaway boxes and plastic straws.

MONTREAL and VICTORIA, CANADA banned single-use plastic bags.

MALIBU, MIAMI, SEATTLE AND FORT MYERS in the US banned plastic cutlery and straws.

2019 TAIWAN banned single-use straws and bags from use in food and drink establishments.

VANCOUVER banned plastic straws and Styrofoam takeaway containers.

NEW ZEALAND banned plastic bags with hefty fines for businesses giving them out.

CALIFORNIA prohibited restaurants from providing plastic straws unless requested.

UN member states agreed to curb single-use plastics by 2030 at the UNEA in Nairobi.

2021 The EU hopes to implement a single-use plastic ban across the bloc by 2030.

Sometimes it feels as if we might just be making a difference. Here at the #2minutebeachclean we've never been busier. Our social media feeds are full of beach cleans and litter picks and we've enjoyed lots of success as early adopters of the cause. Every week we read more and more about the positive efforts people are making to change. Supermarkets make promises, magazines change their wraps, projects are announced that will recycle fishing nets. Our contemporaries, who have struggled for years alongside us, are also enjoying a day in the sun. Lots of new Instagram accounts have been opened in the name of waste-free living. The media now reports on plastic as a constant and major issue, knowing it's in the public eye and therefore 'newsworthy'. The shift of focus has been positive and welcome.

It's also inspiring to see such action on plastics in places like San Francisco International Airport where they have announced a (partial) ban on plastic water bottles. It shows a willingness to change.

The Marine Conservation Society (MCS) has seen a dramatic rise in the numbers of people volunteering for their Autumn Beach Cleans, with 2018 a record year. At the time of writing, it looks like the hugely respected charity and lobbying group will smash their numbers again in 2019. Well done to those who picked up and logged every piece of plastic. Not only does it clear up the mess, but it also informs us what's out there so we can tackle it head on.

AND YET . . .

In November 2018, a whale washed up on a beach in Indonesia with a stomach containing 5.9kg (13lbs) of plastic waste, comprising 115 plastic cups, 4 plastic bottles, 25 plastic bags, 2 flip-flops, a nylon sack and more than 1,000 other assorted pieces of plastic.

There have been a lot of promises made by companies and governments. And yet do you notice that the only tangible actions appear to be coming from us or because of us?

We are the ones using less plastic and we are the ones who have changed our behaviour.

We are the ones opening plastic-free shops and changing our shopping habits. While we might be in the minority, we are the ones driving this thing along.

PLASTIC OFFSETTING

In the aftermath of *Blue Planet II* we became aware of a concept called 'plastic offsetting'. The idea of this is that companies and individuals can contribute to beach cleaning NGOs in the same way that they might purchase carbon credits to offset emissions. The problem with this, in exactly the same way it's a problem with carbon offsetting, is that it doesn't actually solve anything.

We can't carry on producing so much plastic if we're not making plans to dispose of it properly. Paying someone to clean a beach isn't the same as stopping plastic at source. It's treating the symptom, not the cause.

Offsetting means that companies and individuals who pay for beach-cleaning activities or recycling projects can now live apparently 'conscience free' by buying their way out of their plastic habits.

I have a big problem with this because it's an excuse to continue as before. You might as well chuck a bottle out of the window of your car into a river and claim it's OK because you offset your plastic footprint. We don't need that. What we really need is change, not sidestepping. We need you to refill a bottle and stop giving your money to the plastic producers. That's how change happens.

Offsetting is greenwashing.

A RECAP OF WHERE WE ARE TODAY

According to the first comprehensive study of all the plastic ever created, published in July 2017, it is estimated that we have produced around 8,300 million metric tons (Mt) of virgin plastics.

Ninety-one per cent of that plastic wasn't recycled. Twelve per cent was incinerated. Seventy-nine per cent of it accumulated in landfills or in the natural environment.

SERIOUSLY?

Single-use plastic is one of the biggest problems when it comes to ocean plastics. In *No. More. Plastic.* I reported that seven basic items (bottle tops and lids, drinks bottles, drinking straws and cutlery, crisp and sweet wrappers, cotton bud sticks, plastic bags and wet wipes and pads) made up 30 per cent of litter picked up and recorded by the #2minutebeachclean family using our app. You can give up any of these any time, replacing them with an easy switch, and in doing so effectively do away with 30 per cent of beach litter.

SINGLE-USE PLASTIC – BEATING THE HABIT

Single-use plastic perpetuates the myth that we can use something and then just throw it away. The trouble, as we know, is that it doesn't go away. We can't just dispose of our unwanted items and hope that they will disappear. Out of sight, out of mind.

One of the criticisms of the anti-plastic movement has been that people are worried that giving up plastic means having to live without. They worry that they might have to 'go and live in a cave'. Giving up single-use plastic doesn't always mean depriving ourselves. There are lots of work arounds. And while I admit that we might sometimes have to look backwards – to the way society did things before plastic – for better answers, it will enable us to move forwards without wrecking the planet in the process.

If we are to free ourselves from the tyranny of convenience then we will have to learn to slow down, make time for the things that are important to us and give less time to those that aren't.

There is good news when it comes to plastic. But there is also some very bad news. So while we might read about the promises of the big corporations and successes on the beaches, the current plastic story-loving media and climate allows us to also read about plastic in Arctic ice, in snowfall, in our drinking water (it's apparently OK to drink but we need to know more), in every animal tested, in every river and in our children's lungs.

What does it tell us? We are making changes, yes, but it may never be enough, if we continue at this speed. It tells us, surely, that we need to just stop. We need to clean up, very quickly. And we need to find another way, and fast.

Time to Act:

Dealing with waste can sometimes seem overwhelming. But it needn't be. Sometimes the smallest changes can add up to a lot. You might (will) have to put yourself out a little bit to make some changes, and it won't always be the easiest thing to do, but it does matter.

So. Time to reinvent your day. Can you get rid of the following single-use plastic items from your life?

Plastics in your lunch: it's easy to go to the nearest coffee shop and pick up lunch. Who can blame you for that? But if it's important to you to cut out plastic then your lunch is the best place to start. Getting up a little earlier to make a packed lunch will help, as will planning your grocery shop to include items you want for lunch. The side benefits, of course, are that you'll probably save money and you might even get to eat a healthier lunch. You will produce zero waste from what you eat each working day (it's not that difficult) and save approximately 235 bottles, bags of crisps and sandwich packs each year if, for example, you usually go for a meal deal each day.

Smoking: cigarette butts are among the most littered items on earth. They are made from cellulose acetate, a type of plastic, and leach nicotine and all kinds of other nasty chemicals when discarded in the environment. My friend, Dom Ferris, a campaigner at Surfers Against Sewage, reminds us whenever we attend his beach cleans that 'one cigarette butt discarded in the ocean will kill everything in the equivalent of a bucket of water around it'.

Put your butts in an ashtray, carry a tin with you if you can't guarantee it.

Chewing gum: yes, much of it is made with a type of polymer. Surprising? You bet. Years ago we used to chew tree sap. Thankfully, it is now possible to buy plant-based gum. Try Chewsy or Simply Gum.

Vaping: it might be worse than we thought: another clever way of doing the same thing but not, with disastrous consequences. Vaping devices are often single-use and contain batteries, which will leach chemicals if not disposed of properly. They are composite, which means they need to be dismantled before recycling unless sent to a specialist vape recycling scheme like one run by www.blu.com. Read blogs like this: www.plasticexpert.co.uk/recycle-e-cigarettes or ask the manufacturer of your brand what to do with it. Don't just chuck it.

Drinks bottles: carry a water bottle and refill, for free, at thousands of water stations around the UK. And if you want another reason to give up the fizzy drinks, some of them contain up to ten teaspoons of sugar in each single-use bottle!

I talked about the issues with bottled water in *No. More. Plastic.* It's hundreds of times more expensive than tap water, is packaged in plastic (often virgin plastic) and 93 per cent of tested bottles (from a range of brands tested in the USA and Europe) were found to have plastic particles in them. But it seems that human ingenuity is still continuing to get in the way of good sense. Or should I say that the human compulsion to make money is continuing to get in the way of real, human social progress.

Since *Blue Planet II*, we might have turned against the plastic water bottle in droves, which is marvellous. But we somehow still think we're too good for tap water. Instead, we buy water in a can or a Tetra Pak and convince ourselves that we are saving the world on the basis that: aluminium is infinitely recyclable. *Yay*. Tetra Pack is recyclable. *Yay*.

Let's rewind a little and look at what it takes to put water in a can or tetra pack and the waste they produce.

Aluminium cans: these contain more recycled content than plastic bottles (anywhere between 60 and 90 per cent), but still require virgin content to make, a process which involves strip mining bauxite ore. So, while aluminium can be infinitely recycled, it still requires transport and energy to recover and recycle. Recycling rates with aluminium are much higher than plastic and, according to the Aluminium Packaging Recycling Organisation (Alupro), the overall aluminium packaging rate has continued to rise steadily too, from 41 per cent in 2010 to 52 per cent in 2018. In addition, they state that 95 per cent of aluminium packaging collected in the UK is recycled within Europe. So, yes, go for aluminium cans of water, as it takes us a step closer to the circular economy. But only when you forgot your refillable bottle, surely?

Tetra Pak: then there's this 'revolutionary' new way of packaging water. At first glance it looks like a planet friendly solution because of the high paper content (around 70 per cent), the weight (which makes transport costs cheaper) and the more efficient way in which Tetra Paks can be packed (square cartons fit better together than round cans).

However, recycling them isn't as easy as it might seem because they contain paper, aluminium and plastic, making them composite and therefore demanding their own recycling stream and infrastructure. There are few facilities that can recycle them in the UK and, until recently, lots of local authorities wouldn't accept them as part of kerbside recycling. Some still don't, while others only offer recycling at certain locations.

A lack of local recycling facilities was the issue that led to a *Guardian* report in December 2018 about an environmental disaster unfolding in Vietnam. With the country in their sights as a 'new market', Tetra Pak and milk producers pounced, doing deals with the government to flood the market (schools included) with long-life, sweetened milk and yet not doing anything about recovery of the cartons. Tetra Pak were at the forefront of the milk boom in Vietnam, according to the article, educating customers about the convenience and safety of single-use milk cartons and yet producing a whole heap of litter in the process.

Seaweed bottles: water bottles made from seaweed were trialled at the 2019 London marathon as an answer to the problem of thousands of single-use plastic bottles littering the streets after the race was over. The *Metro*, gushing with excitement at this new development said this: 'If runners don't fancy eating their pouch, they can chuck it on the floor – the pods biodegrade within six weeks, which is far more environmentally friendly than plastic's degrading time of hundreds of years.' More environmentally friendly? But still not perfect, when we consider the environmental cost of production, resources and transport. And we know it, right?

We might create all kinds of new ways of packaging and transporting the stuff we can make money from, but sometimes we need to look beyond money and innovative solutions that use energy, take up resources and give us something to share on Facebook. Sometimes we need to look to the past for a better future: welcome back, the humble tap. No transport cost, relatively low cost compared to bottled water (about 500 times less), 99 per cent compliance with health and safety rules and universally available. And also anywhere between 300 and 1,000 times smaller carbon footprint too.

I was involved in the original ReFill campaign, even though it was just a small part (I wrote all the advertising and marketing materials for the first launch), which began in Bude, my hometown, in 2014. Deb Rosser, a local activist, worked with Keep Britain Tidy's BeachCare, fronted by a beach cleaning expert called Neil Hembrow, to set up a local scheme whereby they would raise money for the local Sea Pool charity and stop plastic water bottles from polluting our beach by selling reusable water bottles. They persuaded local cafes and restaurants to allow people to refill their bottles for free if they asked. The scheme was seen by City to Sea founder Natalie Fee who then – rather brilliantly – took it to Bristol and turned it into a city and country-wide scheme. Since then, over 20,000 locations have declared themselves as ReFill stations and many millions of plastic bottles have been saved from going to landfill, energy recovery or the environment. Download the ReFill app at www.refill.org.

When it comes to plastic alternatives, it's either a case of reassessing the need for plastic at all or challenging those that seem too good to be true. Look at the carbon footprint. Look at where they come from. Look at how the raw materials are extracted or grown. If everything stacks up, embrace them when you come into contact with them. Support the companies that use them.

Make the good guys rich instead.

Before we move on, here are some sobering thoughts from the plastic age:

- Plastic has been found to be present in 93 per cent of bottled water, with an average of 325 microplastic particles in each litre of the 11 global brands that were tested.

- Plastics have also been found in tap water, at an average rate of 40 particles per gallon in 72 per cent of European tap water and 93 per cent of tap water in the USA. It's frightening, but a lot less than bottled water!
- One German study also found more plastic particles in glass bottles of water than in PET bottles.
- The World Health Organisation put out a report about plastic in drinking water in August 2019 that said microplastics do not appear to pose a health risk at current levels, but that much more research is needed as study methods were quite weak.

Shampoos: use solid bars of shampoo and conditioner and soap. Squirty soap dispensers are composite, contain metal springs and are difficult to recycle.

Sauces: go for those that are sold in glass bottles and, when it comes to salad dressings, make your own with oil and vinegar or lemon juice. Take your old bottles to buy oil and vinegar at your local refill or waste-free shop. Get creative!

Household cleaning: use cleaning soaps, soapnuts or refill containers from waste-free shops.

Plastic cotton bud sticks: if you flush them down the loo (though who actually does that?) they go down the drain and to the sewage treatment plant where they can escape the screening process, ending up on the beach, breaking down into microplastics and coming back in your food. Use paper-stemmed versions if you can't live without them. Only put them in the bin. I find cotton bud sticks on almost every beach I visit.

Wet wipes and cotton pads: wet wipes, on the whole, are plastic. Don't flush them, even the so-called flushable wipes. They all get clogged up in fatbergs in our sewers. Use a washable flannel and produce ZERO waste from your daily face wash. If you're worried about bacteria use a new flannel every day, then bung them all in the wash together at the end of the week. Lots of feminine hygiene products are plastic too (see the brand section on page 248 for more). You shouldn't be flushing them anyway. When you do, we pick them up off the beach when there is a lot of rain (combined with sewerage overflows, which are sewers that join up with drains and overflows, get released directly into the sea after heavy rain, to avoid flooding).

Plastic bags: in 2016, just after the plastic bag charge was introduced in the UK, the Marine Conservation Society's Autumn Beach Clean surveys noted a drop of almost half in plastic bags picked up by volunteers, proving in some way that this kind of legislation works. But while the number of plastic bags on the beach has decreased, we still find them. So many of them. It might not be your shopping bags any more, which is fab, but we're still finding plastic bags that have been used for lots of other purposes: the bags used to seal your food, those used to seal that thing you bought, the ones that dog poo goes in, the type that your online shopping comes in and the kind that gets used every single day by thousands of businesses that also fall outside of the charge, like butchers or fishmongers (or come from other countries). A lot of plastic bags can be recycled, which is great, but wouldn't it be great if we didn't have to? What if there were none? Use a cotton bag for life for your food shopping, and vegetable bags for your loose veg and create zero waste!

Internet shopping: Many, but not all, internet companies use plastic bags to send out their products to you. For those who object, it's infuriating and causes a headache: it might say on the packaging that it is recyclable but it doesn't always mean it can be recycled locally. In these cases, I am afraid, it might be time to vote with your feet if you want to avoid it in future. It's confusing: some retailers are using reusable bags that can be sent back, others are using new plastics or bio plastics and others are using 'biodegradable' plastics.

Amazon, famously, got in hot water in 2019 when they brought out new Prime packaging that is unrecyclable in many parts of the UK.

There is good news, however: some retailers are working hard to remove plastic packaging, trial returnable and reusable packaging and switch to recycled cardboard. These include companies owned by the Internet Fusion Group, and smaller eco-brands like Howies and Finisterre, among others.

Straws and cutlery: 4.25 per cent of litter picked up on beach cleans and recorded using the #2minutebeachclean app was a plastic straw or piece of single-use plastic cutlery. On a personal level I find plastic straws on almost every #2minutebeachclean. They are ubiquitous.

Following a horrific video of a turtle having a straw removed from its snout, in July 2018 MacDonald's replaced all their plastic straws in their restaurants across the UK with paper versions as a direct result of the outcry against them damaging the environment. This stopped 1.8 million straws a day from going straight to landfill or escaping into the environment. That's 657,000,000 a year (the USA uses 500 million straws a day, incidentally).

And yet ... In spring of 2019, over 51,000 people in the UK signed an online petition to McDonald's to bring back plastic straws because they don't perform as well as paper when sipping on a milkshake. I'm sorry, what?!

In 2019, California Assembly Bill 1884 prohibited full-service (not takeaway) restaurants from providing single-use plastic straws unless requested. But soon after, President Trump introduced his plastic Trump Straws at a staggering $15 for 10. (Yes, he actually introduced his own brand of plastic drinking straws because 'liberal paper straws don't work'.) According to the marketers, Trump netted around $500,000 in the first week. Trump supporters apparently don't associate paper straws with helping the planet but with the opposing party. Straws as political tools. That's mental. Absolutely raving, planet-screwing mental.

Plastic straws can be recycled, but, because of their size, it is very difficult. They get lost in the system. They also get washed down drains, into rivers and out to sea. They do not degrade. Paper straws can tend to get soggy. They are also quite difficult to recycle as they get contaminated with food stuff. But if they get into the environment they break down relatively quickly.

Which one is best? Neither, when most people don't actually need one. But, if, for some reason you do need to drink with a straw, consider a metal or bamboo straw that you can wash and reuse. And when you're out, be assertive when it comes to straws. Say 'no straw thanks' when you order your drink. I find throwaway cutlery absurd, even the PLA and wooden stuff. It's no big deal to stick a few items of reusable cutlery in your work bag is it? Constantly 'on the run'? Would it kill you to stop and sit down and eat properly?

Crisp and sweet wrappers: yes, those pesky old favourites still wash up time and time again. Sweet and crisp wrappers aren't easy to recycle because they often contain foil and plastic. Thankfully, Walkers recently set up a recycling scheme for crisp packets with Terracycle, who also happen to take sweet wrappers. The alternative is to send them straight to landfill.

Try brands with compostable packaging (Two Farmers crisps are tasty and the packets are compostable) or try brands that still use foil and paper (like Lindt). Travel sweets come in tins that can be sent for scrap or used for your odd bits and pieces (like your grandad used to do).

Get a Terracycle box here: www.zerowasteboxes.terracycle.co.uk. Alternatively, go for plastic-free alternatives (see the back of the book) or give them up. (It might be better for you!) If your local council, school, business, community group or pub has to pay for waste to be removed it might be worth doing some simple sums for them: if they pay for a Terracycle box, will it save them money if they currently put all crisp packets and sweet wrappers in the general waste? If it does, suggest it.

Late-night takeaways: How does your kebab come these days? In Styrofoam? It's a bad news/good news scenario I am afraid. Styrofoam is recyclable. But only if taken to a specialist recycler, and not if its contaminated with food.

Likewise with curries and other delicacies. Plastic tubs can be reused, aluminium and paper can be recycled, if clean. If you're served in PLA (compostables) ask the takeaway outlet what they recommend: do they do a take-back scheme? If not, you might have difficult decisions to come. Do you forgo your principles for the sake of a takeaway. Name your price. This is the reality.

If your local takeaway uses plastic tubs with resealable lids, clean them out and use them to take to the deli counter at your supermarket. Fish and chips and kebabs taste just as good in paper. Find a takeaway with reusable or green cartons.

Polystyrene (Styrofoam) takeaway cartons will be banned by the UK government in 2021, but why not stop using them now?

Ready meals: in the UK, frozen food giant Iceland is making huge strides in reducing plastic packaging from their frozen ready meals, vegan ranges and Slimming World ranges. If you struggle with cooking and seek an alternative to reduce your plastic waste, this company may hold the answer.

Nitrous oxide: never mind the health risks of inhaling Britain's favourite laughing product, the cannisters and balloons left over after the party are no laughing matter. N2O canisters are not refillable but can be recycled (as long as they are empty, as full ones can explode), while balloons used to inhale it are highly dangerous to wildlife, even if they are biodegradable. So, you know, if you must, be a good sport and pick it up.

TAKE IT TO THE NEXT LEVEL: WHERE ELSE YOU MIGHT HAVE TO LEARN TO LIVE WITHOUT PLASTIC

Festivals: lots of festivals are plastic-free these days, which is fantastic. But there are still touchpoints to be wary of. You can avoid ANY kind of plastic by taking your own steel pint cup (it needs to be certified as a pint) and reusing it. Cutlery can be an issue too, so take your own camping fork or a spork to reuse. And, of course, don't forget your water bottle.

Glitter and sequins: there has been lots of publicity about this. Ordinary glitter is plastic. There are now lots of bio glitter products available, including cellulose, 90 per cent bio and PLA. Choose carefully as not all will degrade when you wash it off and down the drain, as you might assume. And a lot of it ends up on the ocean. If in doubt, try not to wash it down the sink at the end of the festival. Or just don't use it.

Balloons: the people who make latex balloons will tell you that their balloons will biodegrade 'in about the same time as an oak leaf'. This can be months, by which time the balloon could have killed wildlife. We find balloons, or the knots of balloons, on beach cleans a lot. When they pop they look just like tasty jellyfish. Sorry to be the fun police on this, but balloons, even the biodegradable ones, pose a threat to wildlife when they aren't dealt with properly. Even then they just go to landfill or energy recovery, they are just not safe in the environment.

Hallowe'en: *Edie*, the responsible retail magazine, reported in 2016 that 15 million pumpkins get carved in the UK – none of which are eaten – and around 12,500 tonnes of Hallowe'en costumes get sent to landfill each year. At the very least, learn how to make pumpkin soup. And when it comes to costumes, make your own! Don't spend your money on tat that will be thrown away the next day. There are worse horrors lurking in your wardrobe that could easily be customised with some fake blood.

Christmas: when it comes to single-use plastic, in this country, Christmas is king – from wrapping paper to tinsel, decorations and novelty gifts. Do the scrunch test on wrapping paper. If it stays scrunched it's paper, if it springs back, it's plastic. Make your own cards, watch out for tinsel bits escaping (it's plastic), make your own decorations from paper, cinnamon sticks, dried fruit and cloves (lovely smells). Brown wrapping paper can be personalised with crayons and then reused or recycled or composted.

The British Christmas Tree Growers Association estimates around 7 million trees are bought each year. Plastic trees have a far higher carbon footprint than real trees, but real trees, kept alive from year to year and reused, are by far the best option. Trees absorb carbon, and so will your Christmas tree if you let it live beyond Yuletide. If you dispose of it by composting it or chipping it, it will release the carbon it has sequestered back into the environment eventually. How about making a tree out of branches of clean deadwood or driftwood? Use it every year and keep that carbon stored within the wood.

In 2017, it was estimated that 127,000 tonnes of plastic packaging went to landfill. And that's just what surrounds the presents, many of which I'll wager never get used more than once. Those hilarious nylon socks you got for Uncle Gerald? Your uncle Gerald would much rather have something you made yourself. Fudge is easy, as are biscuits, and they don't create any waste. For kids, buy gift cards for digital goodies like music, games or online subscriptions. Gift cards for experiences – not things – make much more good sense now we know what we know: a National Trust membership, for example, creates NO WASTE! And guess what? You could save money too.

Holidays and days out: there's no reason you can't travel and be green at the same time. Holidays can be a nightmare for single use, simply for the conveniences presented to you, and also because of issues with local drinking water. However, avoiding single-use plastic will make travel easier with a little planning. If you travel by air it'll make security easier if you avoid liquids altogether. Get through security quickly by using solid soaps and shampoos (and even toothpaste) in tins if you travel light. Avoid using hospitality miniatures. On arrival at your hotel think about buying a pint of milk to put in your hotel fridge, if all they have is milk cartons.

Thousands of cheap bodyboards, popped lilos and other pool inflatables get left on British beaches every summer, going straight to landfill. Cheap holiday tat breaks easily (polystyrene boogie boards typically break after just a few uses) and often can't be recycled unless taken by specialist recyclers or schemes. Avoid. Wooden bellyboards last for years, are made locally and are much more fun. Hiring boards and wetsuits saves having to buy cheap stuff and doing without the inflatables is just good sense. Lifeguards hate them.

Taking kids on days out, or simply heading off for the day, doesn't have to cost a fortune or rely on food in rubbish packaging. Unless there are dedicated bins, most likely it will all end up in general waste. Pack a picnic lunch and drinks bottles and save money. Take a reusable coffee cup and create no waste with your coffee. Eat at sit down restaurants and enjoy nutritious food with proper crockery! If you have to eat on the run, choose wisely:

lots of sandwich shops and takeaways are becoming more responsible and we can show our approval with our custom. Just remember: convenience isn't your friend.

National events: after the London Marathon in 2018, Westminster Council cleared up 27,000 single-use plastic bottles. Sporting events are notorious for creating waste, although things are changing, slowly. But how do you get around stadiums not allowing you to take in water bottles or cups? It's tricky. And those isotonic gel capsules? Do they really work? Is the performance advantage gained more important than the problem of the waste they create, that goes straight to landfill? How about eating before the game/match/event rather than during? Carry a collapsible coffee cup to take in with you? Can you train with a carry water bottle if you're a runner? Run with a CamelBak? Waste-free sport requires a little thinking outside the box, like all of life these days.

BECOME PLASTIC CONSCIOUS

I will write about being 'plastic free' later, but for now I'd like you to become plastic conscious. I need you to open your eyes to plastic, to notice it in your life.

Every time you go anywhere, you will be confronted with plastic. At service stations, train stations, airports, ports and bus stations, leisure centres and in the supermarket you'll find it: protecting a pair of new googles, wrapping your ice cream, displaying that Swiss Army Knife you wanted, the tiny plastic seal at the top of the salad dressing, the see-through film on your BLT. Anywhere you buy something or are exposed to material objects you'll find it.

Plastic will be at work too. Your suppliers will use it to wrap bricks, windows, shoes, reports. In some cases it'll be invaluable because it'll reduce wastage or protect objects from damage. In other cases it'll be because that's the way they do it: plastic is cheap, disposable and useful. They shrink wrap new cars in plastic these days. And kitchen units. And boats too. I find it all very confusing.

THE NEW PLASTIC PROBLEM

Hey! We got a new compostable coffee cup.

You can take it away.

It's cool.

Let's share it on social media.

It gets 1 million views.

Then what?

It's the burning question, and it's been bothering me for a while now. There have been a lot of proclamations about compostable, bio-degradable and bio plastic products that we can now use in place of plastic. At first glance they look like brilliant answers to the plastic crisis. We simply swap one for the other and – hey presto – we carry on as before. Genius. No harm done.

But it's not so simple.

THE TROUBLE WITH COMPOSTABLES

So that cool new compostable coffee cup you got from that cool coffee shop? It might not be all that unless it's home compostable and you can deal with it yourself, or the coffee shop offering the miracle cup is operating a take-back scheme.

The *National Trust Magazine* used to come in plastic wrap. This is to keep it contained, along with its paid-for advertising inserts, so that everything arrived with the member intact. It's the same with lots of magazines. In 2018, the National Trust changed their wrap to a certified home compostable one made from corn starch. The crucial thing that the Trust did was to print instructions on how to compost it on the wrap. This helped to educate the membership about the wrap, how it should be disposed of and also how it should not be disposed of. Others who have replaced plastic wraps on all kinds of items, including magazines, have not been so diligent.

What use is a degradable wrap if you don't know how to degrade it? What use is a compostable bag if you can't compost it at home and can't send it to a composting facility?

Just as printing '100 per cent recyclable' on a plastic bag doesn't mean it will get recycled, so 'doing the right thing' with compostables doesn't mean it will get composted.

Compostables, unless they are certified home compostable, have to be composted in industrial composters that provide the right conditions to break down the type of materials being composted. They need moisture, heat, to be turned and bacterial activity. To be certified compostable, an item must fully compost in an industrial facility in fewer than 12 weeks. In landfill – or if they get into the environment – this could be much, much longer. Long enough to cause damage to wildlife in all probability.

In the UK there are just 50 industrial composting sites. Getting your coffee cups to these sites is a challenge as not all home garden waste collections will accept them. Where organisations operate a take-back system whereby the cups get collected and taken to the right kind of facility – we're talking about schools and hospitals, for example – then it's viable. But if you're leaving it to go to landfill it's worse than useless. Some studies have shown that putting compostable cups in landfill creates more methane than standard cups due to the way they break down.

In addition, putting biodegradable or PLA coffee cups into the standard recycling stream can contaminate it, therefore spoiling the whole batch with impurities.

So what do you do? We can't go on using plastic-lined coffee cups. We use 2.5 billion each year in the UK and, of those, only recycle 0.25 per cent. And the new plastic versions – the compostables – aren't much better unless you know they will definitely be composted. Moving to a bamboo keep cup would appear to be an answer. Or is it?

THE BAD NEWS ABOUT BAMBOO

As I was writing this book, a report came out about the use of bamboo coffee cups. You get given them at events, probably have one in your car and don't go anywhere without one so you can easily avoid single-use coffee cups. I'm the same. Bamboo, on the surface, is a product with plenty of eco credentials.

It's fast growing and natural and, in theory, compostable.

The trouble, according to the report, comes when these coffee cups are put under scrutiny and are tested for their composition and contents. Bamboo coffee cups are made by grinding bamboo into a powder or mulch and then adding melamine to remould them. Melamine contains formaldehyde, which, according to the UK government, can lead to irritation in the nose, throat and mouth, can cause burns and ulcers in the stomach or intestines, may also cause chest or abdominal pain, nausea, vomiting, diarrhoea and gastrointestinal tract haemorrhage and kidney failure, and can also cause cancer . . .

According to the report, bamboo cups containing formaldehyde become potentially unstable over 70 degrees centigrade or when microwaved or washed in a dishwasher. The report also concluded that bamboo cups are not particularly compostable and their 'end of life' scenario would be best taken care of by burning in an energy from waste facility.

A supplier of bamboo coffee cups in the UK claims the quantities are not harmful to us. Even so, the report caused a bit of a stir, even in the #2minutebeachclean office, where we currently use bamboo cups. Ours, thankfully, were not made using melamine or formaldehyde, but in our investigations, we discovered they are made with a type of plastic. This means they cannot be composted, as we first thought, and were promised by our supplier.

This is a good example of eco stuff gone wrong – and 'the right thing' turning out to be the wrong thing. Of course you are reducing your single-use coffee cup consumption by carrying a coffee cup, and they can be used time and time again, but are you prepared to accept the potential side effects?

Time to Act:

Revert: choose a mug that's made from a recyclable or biodegradable material. Steel cups are good. Glass, of course, like the cups made by Sol Cup and Keep Cup are benign. Some brands, like Rcup, are repurposed from old, single-use coffee cups and are recyclable.

Restyle yourself: how about sitting in the coffee shop and asking for a mug? If you can find time, and a coffee shop that doesn't use single-use cups for sitting in (many do), then taking a pew and taking five minutes to slow down and remember who you are can be great for both you and the planet. As for coffee shops that don't have any kind of reusable mug: what is that all about?

THE ISSUE WITH BIOPLASTICS

Bio plastics are made out of renewable sources and can be dropped into existing plastic manufacturing infrastructure without any major technology or equipment investments. But they do not biodegrade. Examples are: bio-ethylene, bio-polyethylene (bio-PE), bio-propylene (bio-PP), bio-polyethylene terephthalate (bio-PET). Bioplastics do not leach chemicals into food but, as was suggested in an article from Columbia University in December 2017 called 'The Truth About Bioplastics', can release methane (a powerful greenhouse gas) if left in landfill.

New plastics, which are designed to do the same job as plastic, can be a great thing. Don't get me wrong, the potential of bio plastic mulches in agriculture (these are ground coverings that suppress weeds, conserve water and provide great growing conditions, benefitting the soil and reduce food miles) is huge.

But new plastics, the way I see it, still need to be used thoughtfully, otherwise they just become more waste. In a lot of ways, new plastics, made from non-oil-based sources, are better because they are made from renewable resources. However, creating them may cause other problems, such as using up agricultural land that should be used for food production, transportation and the end of life.

In 2018 Lego produced a (tiny) range of Lego plants made from ethanol from Brazilian sugar cane, a renewable resource. It's better, but not brilliant, because the sugar cane takes land, water, pesticides and chemicals to grow, which may not prove to be that sustainable.

But the big problem that I have with it is that the ethanol-based polymer is still plastic. It'll still persist in the environment if it escapes the Lego box. Likewise, if it gets into the ocean (it's small and can easily get washed down drains) plant-based Lego may not degrade and risks being eaten by sea birds, who really won't care that it's made from oil or sugar cane when they feed it to their chicks.

Lego, on their website, claim that their bricks can be recycled along with the household recycling. But they are so small that it seems unlikely they will ever make it through the Materials Recovery Facility (MRF). And some of the pieces are made from number 7 plastic (which is classified as 'other' plastics and is a catch all for composites because it doesn't fit into any other plastic category) and cannot be recycled easily anyway (see page 255).

At the moment, there is no dedicated waste stream that deals specifically with plant-based plastics, so what happens to it?

Time to Act:

Reinvigorate: if you have buckets of Lego that you need to get rid of, recycle, sell or donate it. Don't put it in your kerbside recycling as there's no guarantee that it'll get recycled. Better to put a smile on someone else's face than send it to landfill.

Rethink: in you are thinking of buying Lego for your kids, buy it second hand. This way you'll be making good use of the stuff that's already out there and not contributing to the 19 billion bricks that are made each year. Better under your feet and up your hoover than sitting in landfill.

PLASTIC ALTERNATIVES

The search for a viable alternative to plastic is ongoing. Here are some of the current materials.

MATERIAL	REPLACES	ADVANTAGES	DISADVANTAGES
PLA (polyactic acid) Made from the sugars in corn starch, cassava or sugarcane.	Traditional plastics	• Biodegradable • Carbon-neutral • Food safe • Can look and behave like polyethylene (plastic films, packing and bottles), polystyrene (plastic cutlery) or polypropylene (packaging)	• Can only be composted above 58 degrees centigrade in industrial composting facilities • Can emit methane in landfill • Needs to be collected and disposed of properly
Drop-in bioplastics: bio-ethylene bio-polyethylene (bio-PE) bio-propylene (bio-PP) bio-polyethylene terephthalate (bio-PET)	Traditional plastics	• Can be 'dropped in' to standard plastic processing • Can be recycled with standard recycling	• Does not biodegrade • Cannot be composted • More expensive than traditional plastic • Emits methane in landfill
OXO biodegradable thermoplastics (chemicals added during the manufacturing process to allow the plastic to photodegrade)	Traditional plastic bags	• Degrades into carbon, biomass and water • Breaks down quicker than normal plastic due to catalysing chemicals	• Unclear as to correct disposal • Degrades up to 18 months, in the right conditions • EU making plans to ban it • Degradation varies significantly depending on conditions determined by temperature, light intensity and moisture. • May give off CO2 in landfill

MATERIAL	REPLACES	ADVANTAGES	DISADVANTAGES
Mushroom (mycelium grown to make into packaging, faux leather and polystyrene alternative)	Packaging, foam, leather	• Can be grown in any shape • Very quick to grow • Biodegradable • Home compostable • Water repellent • Compostable	• Not as quick to manufacture as traditional plastics
Seaweed (processed into bioplastics)	Bottles (trialled at London marathon in 2019), packaging	• Grows quickly without chemicals • Biodegradable (less than 2 months) • Low carbon • Edible • Potential to be zero waste	• Not as durable as plastic for food use

END OF LIFE

Some new plastics behave like plastics, which they have to if they are to replace oil-based plastic. And that could mean disposal, as with plastic, is a challenge. If the plastic you are buying cannot be put into your home compost, can't be taken to your recycling centre and won't be taken back by the producer, it may turn out to be a complete red herring.

As with any product, unless you can guarantee its end of life – then it could be pointless buying into it.

Thinking about the 'End of Life' is a brilliant guiding principle for just about everything you buy, especially when it comes to new plastics. Every product will have some kind of life cycle – starting with what went into making it and finishing with where it will go once you no longer need it.

Get it right and your impact could be minimal. Get it wrong and you'll simply be contributing to the plastic problem.

Time to Act:

Repeat the question: every time you buy anything, ask yourself where does it come from? Where will it go? What happens to it once you have done with it? Can you sell it? Pass it on? Will it compost? Can it be recycled? Will it, realistically, be recycled? Unsure? Avoid.

FOOD AND FOOD WASTE

Whether we like it or not, we are going to have to make some stark choices about our food in the coming years. With an ever-larger population, a lack of useful land and an ocean that has been ravaged to the point of collapse (according to the UN, 90 per cent of fishing stocks are fully exploited, overexploited or depleted) making good food choices is going to become an ever more ethical dilemma.

Include plastics in that, because food and plastics have long since gone together. Plastics make some kinds of food production possible, enable food to be transported globally, make it safer to store and transport and help to reduce food waste. However, now we are at the point that our food has plastic *in it*, and the production, transportation and packaging of food is decimating the oceans and threatening our current existence, we have to take another look.

Feeding everyone in the world is a constant and growing issue. And yet, each year in the EU we produce 88 million tonnes of food waste. According to the Food Foundation, around 795 million people go hungry daily, with poor nutrition causing 45 per cent of deaths among under 5s. That's 3.1 million deaths a year.

Apart from the actual food wasted, Britain's supermarkets also produce around 800,000 tonnes of plastic packaging waste a year. On its own that's scandalous enough, wouldn't you say?

The food industry, together with the supermarkets, produces, or is responsible for, a huge amount of plastic waste. Much of it is packaging that's used to store and transport food and is used just once before being thrown away. But there is another side to what's on our dinner plates: the plastic used in production.

So while you might have unwrapped your groceries at the Tesco checkout in protest at unnecessary plastic packaging (and good on you for doing it), there's often more to your food's back story than just a few plastic cartons. It can make scary reading.

FOOD PACKAGING

Growers, suppliers and sellers use plastic to protect food so that it reaches us in perfect condition. They wrap delicate fruit in plastic so it doesn't get damaged in transit from wherever it was grown, so that we can walk into our local supermarket and see shelf upon shelf of delicious-looking produce, at any time of the year, anywhere in Europe.

Food is premade for us too, to save us the trouble of cooking or even having to learn to cook. That, too, gets packaged up into delicious-looking, easy-to-make meals that we can whack into the microwave and devour in minutes before discarding the packaging it came in.

Some supermarkets (notably Iceland) are committing to reducing their plastics and are now offering frozen ready meals in card or plant-based compostable alternatives.

Food, at times, has to be sold in packaging because of safety regulations. Boris Johnson's ill-informed kipper rant about plastic freezer packs in food (he blamed the EU for making kipper sellers place plastic freezer packs in with sold-by-post kippers) was based on regulations put in place by the hygiene food standards agency in the UK, to ensure that fresh food is refrigerated when it is sent to a customer. It's not just kippers by post, though, that have to comply with food standards.

Until companies can come up with feasible alternatives, we need plastic to make food transportation, hygiene and storage easier and often cheaper. But at what cost?

The area around Almeria in southern Spain – an area of around 135 square miles – is covered with plastic greenhouses where farmers grow a lot of Europe's fruit and vegetables year-round. What was once desert is now a sea of plastic, cast aside when no longer serviceable and replaced with more.

While our conscious selves might baulk at buying tomatoes from South Africa or South America because of the air miles, buying them from Spain may be no better when you take into account the plastic used and discarded to grow it, the transport costs to get it to you and the environmental damage being done by the chemicals used to grow the plants in the first place. Growing in Almeria also can mean poor working conditions and minimal pay of the migrant workers used to harvest the produce.

Time to Act:

Rethink: learn to question where your food comes from, just as you might count calories or sugar content. Can you guarantee its provenance? If not, find food you can count on.

If you can't get the answers at your supermarket (some are now selling ranges of locally grown produce), shop somewhere else. It will help to shop seasonally too, which means you live off what's available locally. Buying produce that is packaging free won't just help you to reduce your plastic footprint, it will also relieve a part of the burden on your local collection or recycling operations.

Shopping locally does take time and it isn't as easy as popping to the local supermarket. I understand that. But it can mean getting produce that's grown locally, is surprisingly inexpensive and will help to reduce food miles and your food's carbon footprint, and the packaging it comes in. Shopping locally will help to make your local area better off too, as the money stays in the local economy.

You might feel insignificant, but if you and 1,000 other families choose to change the way they spend their UK average household budget on food and non-alcoholic drinks of £3151.20, that adds up to £3,151,200. It's significant enough. With a little will, a town of 10,000 families could put as much as £31,512,000 into the local economy – by spending with local farmers, growers, shops and independents – instead of into the pockets of supermarket shareholders.

THE *BLUE PLANET II* EFFECT

Since *Blue Planet II* aired, some retailers have been trialling plastic-free aisles, using alternatives to plastic and have accepted that people now want to bring their own tubs to the supermarket (and don't look at you funny any more when you do).

Organised mass unwraps have caused retailers to rethink their policies, which is incredible. Campaigns like Hugh Fearnley Whittingstall's War on Waste and SAS' 'return to offender' have now reached the boardrooms of the supermarkets.

Well done.

But are the supermarkets doing enough, especially when you think of the scale of consumer support for plastic-free food?

- **Aldi** pledged to make all of its own-label packaging recyclable, reusable or compostable by 2022.
- **Asda** pledged to remove plastic from its stores where possible.
- **Iceland** pledged to eliminate plastic packaging from its own label range by 2023.
- **Iceland** also introduced a plastic bottle recycling scheme in four of its stores across the UK whereby consumers were rewarded with a 10p voucher for every deposit.
- **Lidl** has announced that its £1.50 Too Good to Waste boxes – which contain slightly damaged fruit and vegetables – will be launching across the UK.
- **Sainsburys** has reduced its own brand packaging by 35 per cent since 2005.

- **Tesco** announced that it is launching a trial to remove a selection of plastic-wrapped fruit and vegetables in a bid to cut down on packaging waste.
- **Waitrose** has said it is committed to stop using black plastic packaging for all own label goods and has removed plastic bags from grocery aisles, replacing them with paper.
- **Marks and Spencer** now takes non-recyclable plastic back at all their stores to turn into playground equipment through Terracycle.

This is all good news. But you can still help to push things along with your individual actions.

Time to Act:

Reuse your bags: as a basic minimum, take your own vegetable bags to the supermarket: Veggio bags are reusable, but are plastic.

Consider cloth bags. At the very least, reuse your old plastic bags as many times as you can. The plastic bag charge, which was introduced in the UK in October 2015, is credited with an 86 per cent drop in the numbers of plastic bags given out at supermarket tills. And yet, in the same period, 1.18 billion bags for life – which are made with thicker plastic and cost around 10p – were sold.

What does this mean? It is possible that we aren't getting the message and are actually taking MORE plastic away from the tills than before. While the charge is positive it's not as simple as it might seem. We still need to change our habits

Recycle: take used takeaway tubs to the deli counter.

Reward: thank the shops that offer plastic-free alternatives by shopping with them.

Rebel: unwrap your products at the checkout (politely) and tell the staff you don't want their packaging (nicely) and tell them why (dealing with waste costs millions and damages the planet).

DO YOU KNOW WHAT'S IN YOUR FOOD?

When I started campaigning about plastics around 2009 I feared the worst. I feared that plastics would work their way into our food, risking exposing ourselves to potentially dangerous chemicals such as BPA. This chemical has long been suspected of causing human health problems by interfering with sex hormones. It has also been linked with some cancers. While the long-term effects of plastic ingestion aren't yet known it seems clear that we need to be vigilant.

Frighteningly, it's starting to come to light just how much plastic we are consuming with our food. In a US article from *Environmental Science and Technology Magazine* published in June 2019, the authors estimated that Americans consume, on average, 50,000 particles of microplastic a year in food, with those who drink bottled water consuming around 90,000 additional microplastic particles annually (86,000 more than those who only drink water from the tap).

While there's no evidence to prove that microfibres – which is what we call tiny pieces of man-made and synthetic textiles that are smaller than 5mm – eaten by fish make it into the flesh we eat (although I suspect it's only a matter of time) you may be ingesting plastic if you like filter feeding shellfish like mussels. One study found 90 pieces of microplastics in one portion of European mussels.

Time to Act:

React: it's time to reassess your relationship with shellfish. Mussels are filter feeders and filter water to get their food. That also means filtering everything else that's in the water. A 2018 study from the University of Hull and Brunel University found microplastics in all samples of mussels tested in UK waters and bought from UK supermarkets.

Research: next time you go to the fishmongers, or the fish counter, ask them about the provenance of your fish. Is it farmed? Is it line caught? Is it local? Is it sustainable? All fishing uses plastic lines, nets and boxes. Local, line-caught fish from MCS Certified 'sustainable stocks' is the best option, both from an ethical standpoint as well as a plastic point of view.

FOOD WASTE: THE CUCUMBER CONUNDRUM

This is my favourite kind of plastic dilemma. Whenever I complain about plastic-wrapped cucumbers I hear the retort that it reduces food waste. Cucumbers last up to five times longer once picked when wrapped in plastic than not as, once removed from the plant, they begin to lose their moisture and begin to go limp. Out of the UK growing season, when UK grown cucumbers are not available, wrapping cucumbers in plastic enables them to be transported from further afield and helps to reduce food waste, a growing environmental problem.

In 2018, Morrison's announced that they would stop wrapping cucumbers in plastic during the UK growing season, which is great, but will continue to wrap those that do not originate in the UK, for reasons stated above.

However, the problem with cucumbers – can be solved quite easily if we relearn how to eat local.

Eating fresh, local, seasonal and sustainable food solves a lot of other issues – and not just with cucumbers. The food miles are reduced, you can check on workers' welfare and growing conditions and put money back into your local economy. Local produce doesn't have to travel far, doesn't need packaging and can be traced for provenance. It seems so obvious but isn't easy, of course. Shopping locally takes time, can be, on the surface, more expensive, isn't easy during the late winter/early spring and has a certain stigma attached to it.

To do this we need to have a shift in the way we consume. We need to savour produce when it's at its best. We need to understand what it means to shop and eat local, fresh and seasonal. We need to learn how to cook from scratch, to buy from local growers and to relish the chance to savour the tastes of where we live, so making our local area richer and healthier.

Time to Act:

OK, what's coming next might sound like an elysian fantasy. It might sound expensive or elitist or difficult. But it needn't be, and it isn't out of reach to consider growing some of your own produce, even for people living in flats in cities. You don't have to do it all at once. Vegetables, even organic ones, can be cheap. Start simple. Start small. Start with good intentions and you won't fail.

Rethink how you shop: order a veg box from your local organic grower, if you can. Head to your greengrocer and buy local vegetables for your dinner tonight. If you don't have a greengrocer nearby, try to buy plastic-free, locally grown veg from the veg aisle of the supermarket.

Rebel: vote with your wallet. It's the most powerful form of protest. Walk away from producers you don't trust and go to those you can.

HOW THE WASTE-FREE SHOP MIGHT JUST SAVE THE WORLD

Let's move on to something altogether more positive. I have a waste-free shop near me. It's run by friends of mine. They stock all dry staples, including rice, flour, nuts and pulses as well as oils, laundry liquids and detergents. It's small and friendly and gives us a completely new way of shopping. We put aside some time to go in, allow ourselves to stop and chat with the people we meet in there and buy stuff that would otherwise have come in plastic packets from the supermarket. It's a win–win.

Since it opened, we've been able to reduce our weekly rubbish by more than 75 per cent to little more than a couple of kilos of unrecyclable packets. And we're working on that!

The concept is simple: you take your own containers in. They weigh it and then you go and fill it. They weight it again and you pay for what you have taken. Using old takeaway tubs, plastic bags, Kilner jars or glass bottles is OK. There's no risk of cross contamination and you take as much as you need. The prices are competitive and, as I have already said, you get to enjoy a different experience. It also saves you the chore of having to go to the supermarket so often.

PLASTIC-FREE AISLES

Waitrose began a trial in one of its stores and has since extended it to three other stores. Much like a waste free-shop, it enables customers to take in their own packaging to refill. It shows the power the consumers have had when it comes to influencing policy – and actually getting the supermarkets to move.

Time to Act:

Reconnect: find your local waste-free shop and change the way you shop. It's much more sociable and will enable you to cut your waste significantly.

Find your nearest at: www.zerowastenear.me

Refill: take cotton bags to the supermarket for your fruit and veg. Use Tupperware to refill. Use old plastic bags to refill. Use anything but something new! The more we use plastic-free aisles, the more will appear.

DO WE NEED TO EAT SO MUCH?

According to WRAP (Waste and Resources Action Plan, see page 51), we buy around 41 million tonnes of food in the UK annually. We waste around 7.1 million tonnes of that, meaning that we throw away around a fifth of all the food we buy. That's insane. You might as well throw away two eggs out of every dozen and be done with it. And don't forget that all that wasted food has to be contained in something to get it out of your home – the black bin liner destined for landfill perhaps?

The good news, though, is that 190,000 tonnes of waste food was redistributed between 2015 and 2018, which means we're getting to grips with the issue, though we could still do so much more.

WONKY VEG AND FOOD WASTE

In 2013, ugly veg was blamed for up to 40 per cent of fruit and vegetable waste in the UK. This was partly due to it being discarded in the field for not being up to 'consumer standards' or being rejected in the supermarket because we didn't like the look of it. Some of it was wasted because it passed a sell by date.

I'm a bit ashamed that my veg selection might have such disastrous consequences, to be honest. And I am a bit cross with the supermarkets for sacrificing so much because it didn't look right or didn't 'fit our expectations'.

Happily, since then things have changed and you can now buy 'wonky veg' boxes from Morrisons and Lidl as well as online retailers like Riverford, while others, like Rubies in the Rubble, make a point of making products from out of date or overripe fruit and veg.

While food poverty is unacceptable anywhere, there are lots of charities who redistribute food waste to those who need it and, happily, more and more supermarkets have teamed up with charity partners in recent years for that purpose. In 2016, France made it law that no supermarket should allow food to go to waste if it was still edible after reports that supermarkets were dousing still edible food in bleach to prevent people from foraging for it at night.

Time to Act:

Reheat: leftovers – who remembers those? Curries are way better on day two, as are chillies and sometimes pizza. Freeze it and enjoy on another day.

Redistribute: give someone in need a meal or two and make their day. Consider giving your unopened excess food to a food bank so others might be able to eat better.

Recycle: use a caddy for your cooked and uncooked food waste and send it to compost or to your local council food waste scheme (if they have one). It will either go to energy from waste or to make compost.

Resist: if you have cooked food and need to throw it away, resist the urge to put it in general waste. It costs your council money to send it to landfill that they'd rather be spending on mental health services. Use your food caddy or, better still, don't waste it!

Resize your dinner: cooking smaller portions will ensure you waste less and will save you money in the long run if you can also reduce your food shopping. It could also mean you can spend more on locally grown ingredients.

THE WASTE IT TAKES TO MAKE YOUR FOOD

Agriculture produces around 135,500 tonnes of plastic waste each year in the UK, notwithstanding plastic packaging. That's around 1.5 per cent of the total waste stream in England.

When it comes to plastic wrapping, even organic food can still be a headache. Organic agriculture (and non-organic agriculture to a certain extent) uses plastic in all kinds of ways long before the products get to you.

It seems to be with the deepest irony that organic farming, which enjoys the mythical status of being both better for you, more expensive, free of pesticides and full of anti-oxidants all at the same time, often uses more plastic than conventional farming.

Did you ever consider the use of plastic in organic food production? You may have seen rows and rows of plastic covering fields. These are mulches. They are laid on the field in strips and then pierced to allow plants to grow through. Farmers use plastic mulch for a few reasons: the plastic warms the soil to improve growing conditions like a mini greenhouse. The plastic also retains moisture so the crop doesn't require as much watering. On top of that, the plastic stops pests from being able to develop on the crops and reduces the need for weeding as the sheet suppresses weeds too. As a result, yields are higher.

How do you feel about that? We know that plastic mulches can be recycled but we don't know how much gets recycled globally. An article about mulches in China in July 2019 from Reuters quoted a figure of 2 to 3 million tonnes of plastic mulches used in the country every year, with only 180,000 tonnes of it recycled, meaning the rest is burned, goes to landfill or is ploughed back into the soil, degrading it. The same article said that traces of plastic have been found in Chinese exports of spinach and ginger.

How can you tell if your food has been grown under plastic? Really, unless you contact the grower, there may be no way. Again, buying local gives you a greater chance of finding out the true provenance of your food.

BIODEGRADABLE MULCHES

Starch-based mulches are available and are currently being tested for the effect they can have on the soil. Hopefully a current research project at Coventry University, and others across Europe, will show that these types of mulches are as effective for yield as plastic. If proven to be harmless, and possibly even beneficial, they may even enhance the soil when they get ploughed back into the ground after the harvest – something that may actually save time and effort on the part of the farmer.

Until then, we must rely on farmers to gather, bale and recycle their plastic mulches at the end of the growing season.

Time to Act:

Renew your gardening skills: plant cucumber seeds on your windowsill in the spring and then harvest all summer long.

Revert: how about going back to basics and joining a growing scheme or getting an allotment? Find out where your nearest plots are available at www.gov.uk/apply-allotment.

THE MEAT OF THE MATTER

I'm not vegan or vegetarian by any stretch. However, I do respect the choices of those who are, for ecological, animal welfare and health reasons and I request, politely, that they respect mine.

Despite being an omnivore, I have cut down my meat consumption over the last few years because it seems to be the only right thing to do, for all sorts of reasons. One of the most pressing (for me) is the question of waste. This is the waste created by production, processing, distribution and, of course, by us.

According to the UN, 20 per cent of meat production, which is equal to around 75 million cows, is wasted annually across the globe. As consumers, we are responsible for around a quarter of that waste.

Have we become so detached from our food that we have forgotten where meat comes from and consider it acceptable to throw it away?

My meat consumption is based on standards that align with my lofty intentions for the consumption of vegetables and fruit: if it is reared locally to the highest standards, hasn't taken land away from other forms of more productive agriculture and hasn't been ill-treated or filled with chemicals, I am OK with it. I am lucky enough to live in Cornwall, which makes it easy for me to meet the farmer, buy direct without waste and enjoy produce with the lowest possible food miles.

Wasting good meat is not an option when it's valued properly. If it is produced cheaply, without any thought for climate change, welfare or pollution and is available anywhere, at any time, it loses its value.

Meat eating might still be abhorrent to a growing number of people, and I understand why. However, when it comes to locally reared meat or game, just as with locally grown and seasonal vegetables, the food miles are low, waste is minimal, trust is high and the quality is excellent.

Farmers will argue that rearing dairy or beef herds on forage can actively encourage biodiversity too, indeed some of the world's rarest and most interesting ecosystems – like Scotland's Machair or saltmarshes – can only exist with grazing. Others will talk about the carbon sequestering properties of grasslands – how much carbon dioxide they can absorb and hold on to – and lack of biodiversity in large scale monoculture.

Factory farmed meat requires additional proteins, 40 per cent of which in the UK is soy. According to a Greenpeace report, three quarters of soy used in the UK is grown in Brazil and Argentina, from GM seed, using chemical fertilisers to enhance the depleted soil. The land used to grow the soy fed to our livestock could be used to grow crops for humans instead.

Your choices have an effect on food production, when you add them to everyone else's. Your money will, in some small way, guide production. Choose carefully.

If you must eat meat, save your money for good quality, locally reared cuts instead of gorging on cheap, mass-produced beef with a high carbon footprint. Consider it, like our forefathers did, as a treat, like samphire or strawberries, and savour the flavour. It won't be the cheapest but, if you offset the cost against the saving you'll make from not eating mass produced, cheap factory-reared meat, it's easy to see the benefits.

Time to Act:

Rethink your meat consumption: visit the butcher and buy something special that's produced on a small scale by a local farmer. Avoid the factory farms, ditch the processed burgers, keep away from imports and do not waste a morsel.

Reassess your meat consumption: if you eat a lot of supermarket meat, take stock of how much it costs each week and how much waste it produces in the form of packaging (remember that black food trays go straight to landfill because they can't be identified by sorting machines). Replace it with one or two cuts of local, packaging-free meat from a butcher and compare the cost.

SOMETHING FISHY GOING ON

We can't ignore the fishing industry in the marine plastic crisis, although many do. Fishermen have long had their backs against the wall: fighting quotas, declining stocks and competition from abroad in the form of huge, factory sized ships, and we've been feeling sorry for them. No one wants to tackle the issue properly because it's political, economic and ecological.

But I think it's time. And here it is: if you are serious about saving the ocean from marine plastic you need to seriously consider giving up fish or, at the very least, giving up farmed or industrially caught fish using 'unsustainable' methods.

We are arrogant enough to describe fish as being 'stock', to think that taking fish from the ocean is 'sustainable' and that we have a divine right to feed ourselves on the bounty of the ocean. Now it's coming back to bite us.

Fishing gear, and other detritus from the fishing industry, makes up a large percentage of the plastic in the sea. A survey of the Great Pacific Garbage Patch concluded that at least 46 per cent of it is made up of fishing waste. We are talking nets, traps, floats, line, boxes and associated items. An early analysis of data from the #2minutebeachclean app notes that up to 34 per cent of plastic litter recorded was from fishing.

It is estimated that 10 per cent of the world's ocean plastic pollution is comprised of nets. These are nets that have been lost or discarded at sea. That's around 640,000 tonnes. Unbelievably, nets get discarded at sea to avoid paying for disposal on land, and may get deliberately dropped onto known wrecks to 'mitigate' the damage to sea life.

These nets are known as ghost nets because they keep on fishing, long after they have been lost, killing hundreds of thousands of birds, mammals and fish ever year through entanglement.

In addition, any type of trawling will always catch anything that's bigger than the size of the net's mesh. In extreme cases, dolphins, turtles, seals and sharks get caught in nets and are killed in the process. I have seen the results of this myself: a dead dolphin on the beach with evidence that its tail had been cut off to release it from a net. The Irish Whale and Dolphin Group, at the time, said that there had been a rise in the number of victims washed up, coinciding with the appearance of a number of huge pelagic trawlers in Irish waters.

Trawling, and especially bottom trawling for shrimp, for example, is particularly damaging as it destroys everything in its path and catches everything indiscriminately. The mesh is small, which means undersized and immature fish are caught before they have a chance to breed, so destroying the line. It is estimated that, on average across the world, around 85 per cent of what is hauled in shrimp fisheries nets is bycatch – something other than what the boat set out to catch that is then discarded. In some fisheries this is as high as 98 per cent.

For every couple of shrimp you put on the barbie that's 98 other fish killed.

This is unacceptable. But how do we manage it?

The conclusion? Give up eating factory caught fish, or fish altogether if you can't guarantee its provenance. Plastic waste from fishing is killing the very thing it relies on for its existence.

Give the oceans a chance to recover.

Time to Act:

Research: find out where and how your seafood was caught. If there's no good answer, don't eat it. Use the MCS's Good Fish Guide to sustainable stocks to choose what to eat.

Reward the innovators: in recent times, recovered nets, brought in through the Fishing for Litter scheme or the Odyssey Innovation scheme, are recycled into kayaks. Other schemes turn net into yarn, which is woven into fabric (although how many microfibres they shed due to wear and tear over time is hard to say). Supporting these schemes encourages the recovery of nets and places a value on them. With your support they can do more.

IS FARMED FISH THE FUTURE?

In Europe, fish farming generates 3 billion euros per year and employs an estimated 80,000 people.

The irony with farmed fish is that it still relies on wild fish for food. A report in 2019 called 'Until the Seas Run Dry' by the Changing Markets Foundation and Compassion in World Farming said that almost 70 per cent of landed forage fish (highly nutritious small fish like sardines or anchovy) are processed into fish meal or fish oil for aquaculture. That percentage of landed forage fish represents nearly 20 per cent of the total wild-caught fish landings. In addition, 3–6 million tonnes of low-value fish are captured and used as direct feed, which could amount to 20 per cent of catches in South East Asian countries and up to 50 per cent in Thailand and China.

So we catch wild fish to farm fish? Let's get this straight. Unsaleable small fish (so-called forage fish), sometimes juveniles (that have yet to reproduce) and shellfish, which provide a vital part of the ocean's ecosystems (and feed all the bigger fish) are caught to process into food for farmed fish.

Aquaculture also uses plastic of all types because it is durable, light to handle and strong compared with traditional materials. However, due to bad weather, degradation and accidental loss, aquaculture loses floats, rope, nets, traps and all kinds of its containment infrastructure. Mussel pegs, used to keep mussels on ropes, have washed up regularly on beach cleans near mussel farms, while floats and other large items often turn up where aquaculture is present.

The industry has to clean up to remain credible. It has to minimise losses, recycle what it doesn't need any more, reuse reclaimed gear and stop becoming a source of marine pollution.

And that's not even getting onto the subject of parasites and disease, which is a big problem for some fisheries, in some cases escaping into the wild population, decimating further the stocks of fish that have already been taken for fish food.

Time to Act:

Recover: stop eating any farmed fish. Stop it now. If you want to avoid the inevitable waste that comes from the fishing industry then the only course of action is to stop eating farmed fish and shellfish.

Reel it in: locally landed, line-caught fish is the most sustainable way to eat fish, with the least amount of collateral waste. Food miles are low, stocks and habitats are not decimated by trawling and there is little bycatch. Plastic pollution is also reduced, although there are no figures to back this up. A 2017 report in the

Daily Telegraph quoted a study from Belgium that estimated people who eat shellfish ingest up to 11,000 plastic microfibres a year. The effect on human health is not yet known. But ... you know ... really?

WHAT'S WITH THE WASHING UP?

As we know, your sink has a direct connection to the ocean. In 2019, Friends of the Earth and Bangor University announced the result of a study into microplastics that found them in every lake and waterway they tested. A report from the Plymouth Marine Laboratory also found microplastics in every marine mammal (50 stranded creatures) they studied. Plastics included nylon, fishing net and toothbrush bristles.

Do we need any more reason to give up our old sponge scourers, microfibre cloths and man-made cloths? They are all a part of this, and every time you wash up, wipe down the sink or rinse them out you are contributing in some way.

Time to Act:

Restock: get rid of your sponge scourers and replace with coconut husk scourers: they last well and won't scratch your pots. Lose the microfibre cloths and replace with good old cotton cloths. Replace nylon scrubbing brushes with traditional versions.

Refill: refill bottles of laundry and washing up liquid from a waste-free shop. Buy dishwashing powder that doesn't come in individual sachets.

Retire: soap dispensers might seem like a cleaner choice but they are hard to recycle. Refill them if you can and keep them going for as long as you can, or switch to bars of soap. Soap on a rope, hung from a tap, makes very little mess.

CLOTHING AND TEXTILES

You may not realise it but what you wear can have a great effect on the wellbeing of the planet – whether that's through your choice of fabrics or the way those fabrics are created. There are pros and cons to all fabrics and only you can decide where your limits lie.

While I was doing some research on clothing waste I found a stat that shocked me so much I had to go back and check it a few times. But here it is: each person in the UK has, on average, 57 items of clothing in their wardrobe that don't get worn. That's about 3.6 billion across the nation. One in 20 of us also has over 50 items in our wardrobes with the tags still on.

What?

I find that truly shocking. Not only because we've got to the stage where we have that much that we can stockpile clobber, but also that we've got that much cash that we can spend it on clothes we will never use.

Fast fashion – inexpensive clothing produced rapidly by mass-market retailers in response to the latest trends – is using up our resources. Regular fashion, with its new seasons and constant change, is doing the same thing. Both create mountains of waste that we might not be able to see quite as easily as a pile of old plastic bottles, but that are definitely there.

It's making us poorer too, as WRAP estimates the value of these items at £30 billion with a further £140 billion worth of clothes going to landfill each year.

Should I ask the question again? . . . Are we insane?

FAST FASHION

Fast fashion is all about turnover, getting cheap versions of runway items out onto the high street very quickly. It is the business of selling billions of products, often at a really cheap price and sometimes without any particular concern for worker's welfare or environmental impact.

Fashion, as well as being one of the most polluting industries, uses a lot of man-made fibres – which are made from fossil fuels, of course – to make its clothing so cheaply. It is estimated that petrochemical-based fibres make up 65 per cent of textiles used worldwide.

If made from man-made fibres, fast fashion items will languish in landfill for years (many hundreds by all estimates) or may end up in the oceans or environment, breaking down into microfibres

Not only is fast fashion a waste of resources, it is also costing lives. The garment trade employs between 25 and 60 million people worldwide, making it one of the world's biggest industries. Many of those work for minimum wage, which is often less than enough to survive, and forces them to work longer hours to make up the money for a decent standard of living. Workers cannot rely on unions to help them and are often forced into temporary contracts, giving them no rights.

On 24 April 2013, the Rana Plaza building in Bangladesh, which housed five garment factories, collapsed, killing 1,138 people and injuring another 2,500, most of whom were young women. Despite warnings about cracks in the building's structure the workers were ordered to work on the day of the disaster. Among the fashion brands using the factories were Benetton, Primark and Matalan.

In the years since, thankfully, around 200 major clothing brands signed up to an accord on fire and safety in Bangladesh, which is an agreement between the brands and trade unions representing workers to implement building checks and safety measures that will help to keep workers safe in future.

But is it enough? Is it OK to carry on buying cheap clothes? I would argue that it isn't. Cheap clothes help to ravage the planet, wreck lives and promote the continuation of poor practice.

Time to Act:

Redistribute: don't throw away your unwanted items. Go through your wardrobe and make a pile of the things you no longer wear, haven't worn and are unlikely to wear again. Make sure they are good enough quality to be sold on or worn again and take them to a charity shop.

Refill your wallet: if you have clothes that are still good to wear, sell them on eBay or with the fashion app Depop. If you can keep clothes in circulation for longer you can help to slow down the demand for new clothes as well as stop them from going straight to landfill. And you can make cash along the way.

Repair: if you have clothes you love but that are getting tatty, mend them! Learn to sew or take them to the menders. Repair cafes will help you to keep your clothes alive. Find your nearest at www.repaircafe.org

TIME TO EMBRACE SLOW FASHION?

It's hard to know where to turn for the most ethical choices when it comes to fabric. And it's often a case of having to fork out more for the most ethical choice, which is tough if your budget won't stretch to it.

However, if you buy fewer items you'll have a bigger budget for slow fashion items. These are going to be more expensive but should last longer if you care for them, make your choices carefully and only buy items that you will love and cherish for years to come instead of wearing once and then discarding.

WHY MAN-MADE FIBRES COULD BE A DISASTER FOR THE PLANET

Plastic fibres – nylon, lycra, terylene, polyamide, polyester, acrylic – have helped to make a lot of clothing more affordable. They don't require land to grow (like wool, hemp or cotton) but are made with fossil fuels.

Man-made fibres solve a lot of problems but they have also helped to create a disaster for the oceans.

According to a report from Queensland University of Technology in August 2018, plastic microfibres have now been identified in ecosystems in all regions of the globe. The report estimates that they 'comprise up to 35 per cent of primary microplastic in marine environments, a major proportion of microplastics on coastal shorelines, and to persist for decades in soils treated with sludge from waste water treatment plants.'

In 2016, Eunomia, a research institute based in Bristol, estimated that 190,000 tonnes of microfibres from the production and normal use of synthetic textiles enter the marine environment annually. Included in this figure is domestic laundry.

Plastic particles, including microfibres, have been found in Arctic sea ice, on the sea floor in the arctic and also in Arctic snowfall, suggesting that plastics are now airborne.

Microfibres are shed by all textiles, whether they are clothes, carpets, curtains or washcloths. The problems come when the textiles are made from man-made fibres such as acrylic, polyester, nylon or polyamide because they do not degrade like natural fibres. Instead they persist in the natural environment, fragmenting through wear or by chemical degradation and simply becoming smaller and smaller.

Plastic microfibres that are flushed, washed down drains and sinks or enter the waterways through the drain of washing machines may be filtered by sewage treatment plants (estimates put the figure at around 95 per cent or more that are captured). Those that escape often have a direct route to the ocean via rivers, storm drains or other waterways. These are the bits you are unlikely to see on your beach clean, no matter how hard you look.

On top of that, and perhaps worse still, is sludge from sewage treatment plants is used in Europe and North America as a fertiliser in agriculture, meaning that much of the plastic microfibres that are captured could still end up on our fields and then get washed back down the waterways to the ocean.

Oh, yes, and if you think it's still below panic levels, consider that DEFRA estimates between 160,000 and 3 million homes in the UK have misconnections, sending waste water that should go to a foul sewer (and to a treatment plant) to a surface water system, which outflows into a river system or the sea. Thirty-five per cent of those misconnections are attributed to washing machines. It can be safely assumed that a significant proportion of microfibres that are shed by washing clothes will enter the ocean eventually.

Richard Thompson, of Plymouth University, is the man who coined the phrase microplastics in 2004 and is at the forefront of plastic research. He estimates that up to 700,000 microfibres could be released with every 6kg (13lbs) load of washing. Yep, you read it right.

Time to Act:

While you are unlikely to be able to do anything to clear up the mess, there are a number of things you can do to prevent it getting any worse.

Redress yourself: wearing man-made fibres is a shortcut to sending microfibres down the drain, so, if you can, wear as much cotton, wool, hemp, linen and other natural fibres as possible and you'll minimise the risk. Fibres will still go down the drain, but they will biodegrade quickly and harmlessly.

Restart your wash: if you have gym kit or clothes made from man-made fibres, separate them for washing and wash them separately and less often. If you can, get a microfibre filter for your washing machine, which is either a bag, like a Guppy Bag (see page 250), or a ball like the Coraball, or even an inline filter like a Filtrol. These will contain the plastic microfibres so they can be disposed of safely with your general waste rather than disappear down the drain to the ocean.

Cleaner Seas Group is working on a washing machine filter that will capture 100 per cent of microplastic fibres from washing machines, meaning that you'll be able to continue to wear man-made fibres – and wash them – without worrying about microplastic pollution. Until then, keep using the Guppy Bag and avoid washing your man-mades too often. Also remember to dispose of your clothes responsibly at the end of their life.

Reconnect your home: check that your home is connected to the right sewerage system. Your foul water (toilet, sink, washing machine) should all go to the foul sewer and on to the water treatment plant. Surface drainage, such as water run-off from your roof, should go to the surface water or a combined sewerage system. It will usually be your responsibility to sort this out, unless it's off your property, in which case it'll be your developer or water company who must sort it out. If you suspect a misconnection, contact your local water company.

IS IT SMART TO WEAR RECYCLED PLASTICS?

There's been a lot of noise recently about textiles made from recycled fishing nets and post-consumer waste. It's a great story, fulfils the need for a feel-good finish to ghost gear and would certainly seem to 'close the loop' on the fishing waste. But is it smart to wear clothes that have been made with recovered ocean plastic when you could well be simply putting more ocean plastic back into the ocean, just in smaller pieces each time you wash them? Some would argue not. I would exercise caution. Let's not get overexcited.

Recycled fibres might give us the feelgood factor – because, like buying a bracelet that helps to remove plastic from the ocean you feel you are doing something positive – but be cautious about their use. Is there a danger that they too will shed fibres?

What happens to it after you've finished with it? Question everything.

Surely it's better to turn fishing waste into items that won't break down like kayaks from Odyssey Innovation or skateboards from Bureo?

WHAT ABOUT NATURAL FIBRES?

The cotton industry employs around 250 million people worldwide and helps to feed fast fashion thanks to unsafe or unethical conditions, living standards and pay. Cotton also uses a lot of water, with just one pair of jeans requiring around 10,000 litres of water to grow and make.

There are a number of other choices when it comes to natural fabrics. But each has its own issues.

Here's a very brief overview of the choices and the pitfalls. You decide what's acceptable for you when the time comes to buy new clothes. In the meanwhile, cherish the clothes you already have.

Organic Cotton offers smaller yields than standard cotton, which means it needs more land to grow the same amount and is more expensive, but it does this without the use of pesticides and genetic modification, according to established standards. As with all natural fabrics – and that includes all the fabric on this list – if cotton fibres escape the wash they will biodegrade in the ocean and waterways.

Fairtrade Cotton offers workers a fair price for their cotton and often means that the cotton is also organic (65 per cent of Fairtrade farmers are also certified organic) and GM free. It is often considered the most ethical choice when it comes to cotton.

Wool is often considered one of the most sustainable fabrics but doubts, as always, have been placed over its overall impact on the planet due to its land use and the greenhouse gases emitted by those woolly sheep. There are also questions over animal welfare in some countries. However, it is grown sustainably, is renewable, good to wear and very versatile. And the fibres are 100 per cent biodegradable. It is also one of the most recycled fabrics.

Linen is considered a luxury fabric due to the intensive process required to make it (from flax), but it is extremely versatile. Growing it is considered to be very green too: after it's been harvested, root remnants fertilise and clean the soil, thereby improving its productivity. Growing flax also requires no irrigation, fertilisers or herbicides and pesticides. Seventy per cent of linen grown is in Europe, where it is subject to stricter controls than elsewhere.

Hemp is a massively versatile – and underused – fabric that has suffered greatly due to its association with drugs. It was illegal to grow it in the UK until 1993. Hemp uses much less water than cotton (about 5 per cent of the volume) and is easier to grow, using fewer pesticides. However, it is quite intensive to process in terms of labour. But it also grows much quicker, and almost anywhere, so more of it can be produced using less land.

Bamboo is fast growing, self-renewing and anti-bacterial and is a great fabric to wear. It needs few pesticides. The majority of it is grown in China, which may mean there are few restrictions on the use of pesticides and no information on the eco-credentials of the way it's grown. It is also processed using the same intensive chemical processes that are used to produce Rayon. Possibly not the wonder material it's often being touted as.

Rayon, like **viscose**, is made from wood pulp that is broken down using chemical processes and then spun. While the fabrics are often touted as being natural and eco-friendly, the process uses harmful chemicals, much of which are lost during the process.

Tencel is another fabric that's made from wood pulp but the difference is that the fibre is made with a closed loop process that means the chemicals are used over and over again, making it potentially more eco-friendly that rayon, viscose or bamboo.

Fur – I don't think there's much to say about killing animals just for their pelt.

Fake fur – From an environmental point of view, fake fur is generally made of plastic fibres, which means the fibres it sheds will not biodegrade.

Vegan leather might not be made of animal skins, and therefore more ethical for some, but is often made from PVC or PU, which are both plastics. Their production comes from oil.

Plus other alternatives are emerging on the market these days, made from cork, waxed cotton, paper, pineapples, wine, coffee and even teak leaves.

Time to Act:

Return it: Rapa Nui, a clothing brand on the Isle of Wight, make t-shirts from old t-shirts their customers send back to them (for free). It's a circular process that means less raw material, water and land is taken up than is necessary. Check them out: www.rapanuiclothing.com

Relook at labels: choosing clothes based on look alone is no longer enough for the planet. Looking at the label, and then making a decision based on the type of material used, has to become a vital part of the buying process. Can you live with microfibres going down the drain? Are you comfortable with what will happen to your clothes after you have finished with them?

Rejuvenate your look: you don't have to fork out on high fashion to look good for a special occasion. You can rent luxury outfits from a number of designer rental services to guarantee that one-off look for a party or event. And you'll save a bundle. Try www.nothing-to-wear.com

Reuse: how about gathering up your local prom dresses and running a local prom library? Some communities are taking the pressure off by donating dresses to cash strapped teens. Great idea! Try hire.girlmeetsdress.com

Reclaim: some designers are making high fashion items out of found materials, so reducing the footprint of your favourite stuff. Brands like Urban Outfitters and ASOS Reclaimed have long been upcycling old stuff, while just about every charity shop in Britain has been doing it forever. See www.lindathomasecodesign.co.uk

Revive lost skills: learn how to sew. The longer you keep your clothes, the better they are for the environment. Sewing up holes, patching them and making them last longer will improve their eco credentials no end.

Relearn: wash your nylon, polyester or man-made clothes less often. The less often you wash them, the fewer fibres will make their way to the sea. If you can, wash them in a Guppy Bag (see page 250).

Refresh: hand your old clothes down to a friend or relative, send them to a charity shop or jumble sale or make new clothes from them, if it means you'll love them for longer.

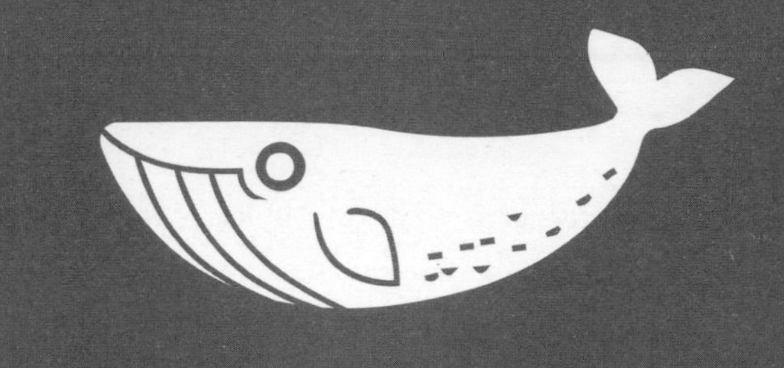

ELECTRICAL AND ELECTRONIC EQUIPMENT

Everything we tap, swipe, watch, listen to, load with laundry or plug in will become waste at some point or another. Some of it will go sooner rather than later, while old favourites, the well-built stuff, will plod on for years, undamaged by the ravages of time. Eventually, though almost everything will give up the ghost. When it does, it has to go somewhere. No matter if it's a toaster that's burnt its final slice, a phone that's too clunky, a plastic power cord with a loose connection, a burnt-out kettle or an out-of-fashion Billy Bass singing fish.

Our society relies on growth and progress. That's the constant evolution of ideas, products and technology to improve our world.

We're clever and can achieve great things. The only thing that we haven't done very well is taking care of the aftermath of all this progress. We leave mountains of waste – old TVs, video recorders, cassette players and piles of plastic vinyl records – behind us in our wake as we seek out the latest thing.

Unless it's dealt with properly it will persist in the environment. It's funny, isn't it, but your old mobile phone, that you used for a year and then upgraded, because the maker brought out a new model or your phone company offered you an early upgrade, has the potential to outlast you by hundreds of years?

WHY WE SHOULD BE RECYCLING

In the first quarter of 2019, the UK Environment Agency reported that almost 323,000 tonnes of household electrical and electronic equipment (EEE) was placed on the UK market by members of producer compliance schemes (producers of EEE must register to a scheme to be able to sell EEE in the UK).

In the same quarter, 123,500 tonnes of household waste electrical electronic equipment (WEEE) was collected by members of producer compliance schemes, according to the EA.

WEEE, which includes computers, TVs, fridges and mobile phones, is becoming one of the fastest growing waste streams in the EU. According the European Commission, 9 million tonnes of WEEE was generated in 2005, with 12 million expected by 2020.

In addition to the plastic contained within your discarded electricals, which you know is problematic, WEEE also contain harmful, rare and valuable materials that MUST be recycled. In some cases it is unsafe to allow those materials to get into the environment. In every other case those materials should be used for other things, because it's a valuable resource and we shouldn't squander it.

For example, depending on who you believe, about 40 mobile phones will together contain a gram of gold. A tonne of phones could contain as much as 300g of gold. It's worth around £40 per gram.

The European WEEE Directive came into force in February 2003 and was then amended in February 2014. The directive requires producers to minimise waste arising from their products and promote their reuse, ensure waste products are treated correctly and meet recovery and recycling targets, and design products to reduce material use and enhance reusability and recyclability. The overall idea is to encourage recycling and the circular economy, preserve resources and ensure they get reused and recirculated.

Another directive, the RoHS (Restriction of Hazardous Substances) Directive, requires producers to limit the use of harmful materials and chemicals, such as lead, mercury, cadmium and hexavalent chromium and flame retardants such as polybrominated biphenyls (PBB) or polybrominated diphenyl ethers (PBDE) in their products.

So much for the EU, eh? What did it ever do for us?

Time to Act:

Rethink: need to replace an electrical item? Think twice before you do. Can it be repaired? Can it be sold? If you buy a replacement, buy second hand or refurbished. Lots of electrical items have a resale value, either on eBay or similar. That way you won't be contributing to the industry, just to the recycling industry.

Return it: under WEEE, producers must take back electricals for disposal free of charge on a like-for-like basis when you buy new equipment. They will dispose of it for you at an approved centre. If you have an item delivered then the retailer must also take your old item away for disposal.

Revive it: before you ditch it, take it to a repair cafe if you can't repair it yourself. Repair cafes are popping up all over the place these days, with volunteers repairing your old electricals, often for free. Find your local café at www.repaircafe.org.

HOW PLANNED OBSOLESCENCE FUELS THE PLASTIC CRISIS

Imagine making something so it breaks or is no longer useful after a certain period of time. That's what is often called 'planned obsolescence'. It's a term that was first coined by General Motors in the USA in the 1920s. The car market was saturated with Model T Fords and new car sales were on the wane. GM brought out a 'new' model, essentially little more than an upgrade to a previous model, and then worked its marketing magic to convince the public that the old one was no longer desirable and the new one was the must-have.

It's a bit like the fashion industry, which brings out a new collection every quarter and then makes you feel uncool if you don't buy it.

Planned obsolescence is a tragic consequence of consumerism and describes a product that is designed to become useless after a period of time or that has a planned demise through a model upgrade (or seasons). For electrical products, this could mean making them with a weak – or cheap – component that renders the item useless and is more expensive to fix than replace; or it could be as simple as creating a desire around the latest version of something that diminishes the worth of the item you already have.

In business terms this makes perfect sense: if you sell a product to a million people and can then slightly update the product sometime later and sell it to the same million people again, then you won't need to find another million new customers. Why would you offer to fix a product's flaws or weaknesses if you can sell them another one?

For example, let's look at the Lightbulb Scam. The centennial light is a lightbulb that was installed in a fire station in Livermore, California in 1901. It was so well made that it still works, over 118 years later. You can watch it glow by logging on to www.centennialbulb.org. When introduced, lightbulbs were originally made with carbon filaments, not tungsten, and they lasted longer. However, in 1924, after electrical lighting had become mainstream, a group of manufacturers, including Osram, Associated Electrical Industries and GE, agreed during a meeting to divide the market up into regions and to reduce a lightbulb's lifespan from 2,000 to 1,000 hours. Customers would always need new bulbs, or at least more of them.

Today it's really no different. Only it's not always the fault of the manufacturers. Products have become so cheap that it's no longer economically viable to mend them. You can't buy spare parts so you simply replace it and chuck the old one away, whether it's a kettle or a monitor, MP3 player or Betamax video recorder.

HOW WE ENCOURAGE OBSOLESCENCE

We help to create the mountains of waste by being seduced by the new, the cheap, the updated and the fashionable. We fall for the rhetoric hook, line and sinker and buy products in our droves. We fuel the demand for smaller, faster, cheaper or more heavily featured products and spend our money on those that fit the bill. Naturally it renders everything that went before it useless and unwanted. And now we have to deal with it.

If something is cheap, it is more desirable and more available to a larger group of people. However, cheapness comes at a price, which is usually quality. Cheaper components get used (plastic instead of metal for example) and products are made to less robust designs in order to keep the costs down. When you're talking about making millions of units, a reduction in cost of just 50 pence per item can make a huge difference to profits, and whether or not customers will choose that product over another.

Not only do these products not last, if they are too cheap they don't hold any value for us. So they become easy for us to discard and replace with another model or something newer. Easy come, easy go. But where is it going?

HOW INNOVATION FUELS THE WASTE CRISIS

In 2000, 2.56 billion CDs were sold around the world. As a composite, CDs are part polycarbonate plastic and part aluminium and therefore defined as number 7 plastic (see chart on page 255). Number 7 plastic cannot be recycled with kerbside recycling and few places accept anything less than industrial quantities. It is technically recyclable, but it's not easy to recycle.

Before CDs there were records and cassette tapes. They are both made from plastics and also difficult to recycle unless they get sent to specialist recycling streams.

Each one represents a stage in the process of enjoying personal music. We've all got hundreds of CDs knocking about now and a whole lot of plastic that will need to be disposed of at some point and already has been discarded now CDs are not trendy any more – they have joined cassettes and record players languishing in landfill, as well as in the loft or in charity shops.

Today our music is digitised, with no waste required to play video or music. It's a great thing when you consider the volume of plastic music that has been produced in the last 50 years. All you need now is power. But then, have you got the right power cord? (And how is that power being generated?) Apple have famously changed their power cord standards for their iPhones and iPods twice. They are switched on when it comes to innovation, but maybe not so much when it comes to waste.

Back in 2016, Apple ditched the 3.5mm jack for their iPhone 7 in favour of the Lightning connector. What it meant was that you couldn't use your Apple headphones unless you purchased either an adaptor or Apple's new USB-C earbuds or wanted to spend £150 on a set of Bluetooth Air buds.

Meanwhile, all your old headphones and connector cables get put to the back of a drawer where they will, very likely, stay and never get used again. Planned obsolescence. So what can you do?

Time to Act

Relisten: got a collection of CDs lying unloved somewhere? Digitise them, store them on a hard drive and then give them to charity if you are planning on throwing them away. Don't put them in the bin and send them to landfill.

Reassess: don't be so quick to jump on the latest this or that. Trends come and go and buying into them doesn't make you happier, more popular, sexier or better. It just means you've got more stuff. When you do need to buy something, ask yourself where it will go once you are done with it – what's the end life? Do you know? Then don't buy it.

Recycle: recylers Terracycle will take all storage media and recycle it. For a price. A small box, at the time of writing, which will take 390 units (we're talking tapes, CDs, videos and cassettes here) will cost around £150 to buy and send back to Terracycle.

Recirculate: components and materials are a valuable resource so if you are not using something and don't think you will in the future, take it to your local recycling centre or donate it to charity.

Rethink: your phone is an environmental disaster. And so are the ones languishing in your drawers. They take precious minerals to make, often require mining that destroys large areas of land and have a dreadful carbon footprint, both in manufacture and usage. They also include plastic and often can't be recycled for their raw materials. With this in mind, the greenest phone you can own is the one in your pocket. So hold off upgrading until you absolutely have to. Take care of your current phone. Look after it. Love it. Buy a case. Don't be fooled by advertising, an offer of an upgrade or the thrill of the new. You don't need it. The average life span of a phone is two years so keep it for longer than this and you're doing a small bit of good.

Redeploy: when you have to upgrade your phone, don't let it lie in a drawer. Sell it, pass it on or donate it to someone who might need it, or recycle it responsibly (do not put it in your recycling bin bag).

Repair it: buy a phone that can easily be repaired, is made with recycled materials where possible and has been created by people living and working in fair conditions. If you don't know the working conditions of the person who made your phone, should you buy it?

The Fairphone 2, winner of multiple awards for green-ness, is home repairable, recyclable and made with conflict-free and Fairtrade materials – shop.fairphone.com.

SHOPPING FOR STUFF

We've already talked about fast fashion and food. But something we've not touched on particularly is packaging, like blister display packs – a moulded plastic shell containing the product, sometimes stuck to carboard – that are specifically designed to display products so you'll want them more. Merchandising, the art of display, enables businesses to easily make their products more visible to the consumer by hanging them on wall displays. Jokes aside about being able to get into them, display packs create a lot of unnecessary waste. I bought a bottle of ink in a bottle to use in my fountain pen. Guess what it came in? A display pack. It's tough to beat the display pack –

they are everywhere. But by becoming aware you can start to see who uses them and who doesn't. Smaller independent shops that have a more traditional selling style might be a good place to start. It does mean taking a trip to the high street – if you have one! – rather than going to the big chain stores.

INTERNET SHOPPING FOR STUFF

In theory, internet shopping should be good for the planet as it means there may be fewer cars on the road driving to the shops and back. But things are never simple. Internet shopping accounts for 22 per cent of all van emissions in the UK, with vans an ever-increasing market that currently accounts for 4 per cent of all of the UK's carbon emissions.

Internet shopping also creates a huge amount of waste, much of it plastic. Bags are cheap and convenient, but, no matter how recyclable or reusable they are, many millions end up in landfill. It's the same with boxes that are too big for the items they contain, with plenty of space and materials wasted. In November 2018, internet sales as a percentage of retail sales peaked at a new high of 21.5 per cent.

I guess this is great for online retailers but it puts an enormous amount of pressure on the recycling system and local waste collections. While cardboard is a standard material put out for collection, only one in ten local authorities collect film, which includes plastic bags, despite the bags being recyclable or recycled. It doesn't really matter what it's made of if it doesn't go to the right place. And the thing about internet shopping is that you have no control over how it comes to you – I have had people tweeting me pictures of my book *No. More. Plastic.* ACTUALLY SHRINK WRAPPED IN PLASTIC!

Internet shopping makes buying possible for lots of people who live in rural areas or can't get out. That's fine. But it doesn't get away from the unwanted waste. Making careful choices when it comes to retailers is vitally important in the battle against unnecessary waste. Some companies, like Surfdome, use packaging machines to pack boxes with minimal waste and don't use plastic bags. If you have a favourite internet retailer that sends their stuff out in plastic, email them. Ask for a plastic-free delivery and allow your voice to be heard. If they don't respond, try another retailer.

Another idea? Shop locally, if you can, and ask yourself if you really need it anyway.

USING YOUR VOICE

Until now, this book has been about taking action on a personal level to deal with the plastic crisis. Making small, sometimes awkward decisions in your everyday life that, cumulatively, can and will add up to make a difference. I firmly believe that every action gets noticed at some level and will have a knock-on effect at some other level, even if its imperceptible at first.

Putting out an increasingly smaller bag of rubbish outside your house on bin day may, subconsciously, have an effect on your neighbours who put out five full bags each week. They might feel they need to do something too and you may notice – with a little help from you perhaps – that it starts to reduce. Even if it doesn't, then your rubbish reduction means 9 fewer bags in landfill every fortnight, 18 each month and 216 each year than it might have been. It all adds up.

And that's what this chapter is about: the power you have to influence the society you live in. Using your voice to make a difference might seem like a small, insignificant act, but you will get heard. And the more you shout, the more you will be listened to.

Your voice is very powerful.

If you buy a takeaway coffee twice a day every day you go to work, on average, you'll avoid sending around 490 cups to landfill each year if you carry a keep cup. Over a ten-year period, that's a fair few (4,900). If you manage to persuade your 99 colleagues to do the same, you'll save almost half a million cups from going to waste, where they will sit in a hole in the ground for a very long time.

Using your influence and voice to inspire the smallest actions – persuading your friends and neighbours to make one small change – can add up to make a big difference.

Just a few years ago, when the #2minutebeachclean was very young, it felt as if we were hitting our heads against a brick wall with plastic waste. It took five years to get any kind of significant funding to help us pick plastic off the beaches. Often we funded the project ourselves, by paying phone bills ourselves or financing travel to beach cleans and events from our own money. My own business suffered as a result because I took time out to go to meetings, attend beach cleans, get our beach clean stations off the ground and to try to get people to listen – because I felt it was important.

There were times when I felt like giving up and handing it all over to some other organisation or forgetting it altogether. What always stopped me was the glimmer of hope that progress was being made. Every tiny step took us closer to the dream of inspiring anyone and everyone to take two minutes out to make our planet nicer.

You can do this too.

I'm not asking you to lose your livelihood or put everything on hold to campaign about plastic and the planet. But I am asking you to give a few minutes of your time to use your voice to add to everyone else's. When the sound of all our voices becomes a roar then things will really begin to happen.

Nowadays, at #2minutebeachclean HQ we get enquiries from all kinds of people all over the world. We get people wanting to set up #2minutebeachclean in their country. People take on personal challenges to help fund us. We've been astounded by the support we get from devoted planet lovers.

We also get contacted by different businesses. Sometimes they want to do a beach clean to offset their practices, at other times they want something for nothing or want the PR from being associated with a green organisation. But more often than not they genuinely want to change. That's when we love to be able to help them reduce their plastic footprint.

What we find out from the businesses that we talk to, is this:

- They want to do something, whether cynically or not, to 'help the oceans'.
- Lots of them think the answer lies in organising beach cleans rather than tackling the supply chain upstream.
- They want to do the right thing but are caught up in a system that relies on plastic.
- Despite the work behind the scenes, they may still give out disposable cups in their canteen.
- They take ages to make changes.
- They are terrified of losing money.
- We can't change their business models.
- They often don't know where to start internally.
- They are getting inundated with complaints about the plastic they produce.
- They read your letters and emails.
- They want to make changes.
- They need our help.

Fighting plastic is tangible. You can see, when you make changes, that your waste reduces. You can also see the destruction that it causes.

Brands realise change is in the air. And this is all because of you, of us, of the power of the 'consumer'.

We still have to stop the flow of plastics into the ocean. We have to learn to live with less and create less waste. We have to return, to some extent, to a local economy. We have to buy products that last. We have to give our money to the good guys. We have to stop giving our money to the incumbent, planet-destroying monoliths. We have to start becoming the change we want to see in the world.

How do we do it? We use our greatest weapons: our money and our voices.

Time to Act:

Refuse: organise a mass unwrap at your local supermarket to prove how much packaging they force on us. Or just unwrap it yourself. Cashiers are used to this these days and will happily take it off you (most of them anyway). They will often be sympathetic too. Surfers Against Sewage organise mass unwraps as part of their Plastic Free initiatives: www.sas.org.

Redeploy your cash: spending money with the companies who do good enables them to do more good. Make the good guys rich by approving of their products. Buying products that have recycled content in their packaging over products that don't, for example, pushes up the value of the recycled material because you increase the demand.

Taking your money away from the 'bad' guys will force them to change. So, while unwrapping food at the till sends a very strong message to supermarkets, for example, taking your money away from them will hurt them more.

Refill: have you got a refill shop near you? Why not start one? It's exactly what hundreds of people just like you have done all over the UK. From Bude to Tynemouth, people have been taking back control from the supermarkets by starting their own waste-free shops. There is a very good network of people to call on for help who have already made the mistakes so you don't have to. Crowdfunding could help you to get the startup costs together to open a waste-free shop: Bude's ReFill shop raised over £12k in just 55 days. Get advice and help about opening a waste-free shop here: www.beunpackaged.com/open-a-zw-shop.

If you own a cafe or restaurant, you can easily become a refill point by offering free tap water refills to anyone who wants it. Or you could rep for the nationwide scheme Refill and talk to cafes, restaurants, bars and shops in your area about becoming a Refill point. And you can help them get rid of plastic straws, stirrers, milk cartons, sauce sachets and sugar sachets too. Join the Refill Revolution here: www.refill.org.uk/add-refill-station.

Return to the high street: the more you shop local, the more the local shops will be able to have fresh, local produce for you to buy. Since *Blue Planet II*, my local deli has been able to improve the quality of its fruit and veg. They sell it unwrapped so more people are buying it, which means the turnover is better, which means less waste and better quality veg all round. If you live in south east England you can find local food here: www.localfoodbritain.com.

Recycle: become a Terracycle recycle point administrator. Terracycle take hard to recycle items and turn them into all kinds of stuff. They run a scheme whereby you can set up local collection points for the general public – perhaps at a school, clubhouse or even a supermarket. Find out more here: www.

terracycle.com/en-GB/about-terracycle/drop_off_locations.

Reassess your choices: set up a bulk buying group and invest in social enterprise or charitable projects. This is easy. Buying in bulk is a great way to save on packaging and save money too. The idea is that you pool your resources to get a better deal. Get a wholesale card, buy in bulk and share it out to save lots of cash (and packaging).

A good place to start might be with Who Gives a Crap toilet paper (or similar), which is not wrapped in plastic and donates 50 per cent of profits to building toilets around the world. Each order of 20 boxes saves 240 plastic wrappers and saves £200 on the cost of buying it individually.

The government offers guidelines on setting up a buying group for not-for-profits: www.gov.uk/government/publications/guide-for-community-buying-groups.

Reconnect with politics: local councils, parish councils and district councils need people to run them and have a say in how your local area works. And they also need people who are prepared to act for the greater good rather than self-interest. It's a big commitment to become a councillor but it can have a big effect. As a councillor you will be able to vote on matters that are important to you. For example, you could campaign to have all local events declared plastic free, putting the onus on the organisers to ban plastic cups, straws, bottles and other single-use items. Government guidance on becoming a local councillor is here: www.gov.uk/government/get-involved/take-part/become-a-councillor.

Resist the litter bugs: set up or join a litter-picking group locally and tackle grot spots in your area. Getting together with other like-minded people will make litter picking light work. Join an online community of litter pickers here: www.litteraction.org.uk. Join an MCS beach clean and help survey the coast www.mcsuk.org/beachwatch/greatbritishbeachclean.

Represent: local councils have meetings every month or so, which you are entitled to attend, ask questions and even video or blog from. Different councils have different protocols when it comes to asking questions at meetings so go online to find out the best way to do this in advance. Find your local council at: www.gov.uk/find-local-council.

Return it: if you are not happy with the packaging your products came in, hate the fact that your favourite company is using plastic or are sick of getting given plastic, even when you don't ask for it, send it back. In 2018, after a campaign to get Walkers to develop recyclable crisp packets inviting crisp lovers everywhere to send their empty packets back to the company, the company announced a crisp packet recycling scheme, which has seen 500,000 crisp packets recycled. During the height of the campaign, Royal Mail were overwhelmed with the response and begged people to put the packets in envelopes as the packets themselves were causing problems with their machines. Send it back to Walkers here: walkers.co.uk/recycle.

If you find litter on the street or beach or in your park send it back to the producer, explaining that it is their responsibility as the producer to ensure their packaging is dealt with responsibly by their customers or is, at least, recyclable. You could also add that you believe they should add bigger messages about recycling and disposal methods on their packaging to encourage better disposal of the product. If you can find a freepost address for the company then you won't have to pay for the postage either. Read more about the Surfers Against Sewage Return to Offenders campaign: www.sas.org.uk/campaign/return-to-offender.

Reclaim your right to complain: your phone is a great resource that can easily be used for good. How? By allowing it to connect you with all the organisations that are doing good. Join Twitter. Join Instagram. Get involved.

Your phone is your mouthpiece and can connect you with people, brands, campaigners and NGOs who are working to rid the world of plastic. It can also connect you directly to the polluters. Give them hell and don't let up. As long as what you do leads to action it's not about being a snowflake (a term that really pisses me off), it's about using the right tools for the job.

- Pick up litter.
- Take a picture of it.
- Post it to social media.

- Tag the people who you think should know. Don't be afraid to call out the polluter or people who are allowing this to happen by adding their @ handle to the post.
- Add hashtags like #2minutebeachclean or #2minutelitterpick by all means – and we'll be glad to see it – but letting the producer know they have failed in their responsibility is vitally important.

Your voice will add to all the others and will be heard. That's the great thing about social media.

Reserve your right to have a say: if you are concerned about a product, a company's conduct or poor record of producer responsibility, let them know. We know that companies are getting a lot of post about plastic but it doesn't mean it should stop. Let them know you want change. Let them know you need to see it happen instead of promises. Tell them you will stop buying their products unless they make changes. You can get the details of any UK company director from Companies House here: beta.companieshouse.gov.uk.

If you aren't getting the kind of answers you want from your least favourite polluter, then you could always find a friendly shareholder to ask questions on your behalf at their annual general meeting. Failing that, buy shares in the company yourself. Then you can demand action. However, while it might sound simple, it may not be, sadly, with some companies requiring you to own a minimum number of shares for automatic qualification to attend AGMs. It may also depend on how you buy the shares.

But if you *can* attend a meeting it's a great way to confront directors directly. They won't be able to get away, at least.

Write to your MP: if you are concerned about plastics, pollution or the speed at which the UK is acting on plastic, write to your MP and tell them. Invite them to come on a beach clean or litter pick (they love that stuff) so they can see what's washing up or being tossed out of car windows. Invite them to your local supermarket to show them all the stuff that can't be recycled or is pointless packaging. Get them to act, as you are.

Don't allow them to fob you off with claims or promises ('The UK has a great record on plastic, in fact it's one of the best in the world'). If you think it's not enough, tell them. If you think they are wasting their time on things that are less important, tell them. Let them know you will withdraw your support unless they act. They have to listen. It's their job.

Remember that you need action points so don't forget to include:

- Tell them what the issue is.
- Tell them why it is an issue for you.

- Tell them what you would like them to do about it.
- Demand action and specify what that action should be.
- Tell them you will be writing again if they do not respond satisfactorily.

And never forget that they work FOR YOU.

YOUR WASTE AND CLIMATE CHANGE

Ever since the #2minutebeachclean became a thing in 2013, I've been wondering how big the plastic crisis would become. First I worried about how plastic affected my local beach, tourism and our health and wellbeing locally. Next I began to worry about how plastic affects wildlife. Then I worried about how plastic affects our food. Now the biggest worry is how plastic affects our climate.

Once you understand the scale of it, you can start to put the pieces together to make up the bigger picture. And that's when things start to get more than a little frightening. If you're anything like me, however, you won't give up. You'll allow this information to fuel the fire of resolve, because to do anything else would be to admit defeat.

THE BIGGER PICTURE

- Plastic is just one small part of our planet's problems.
- Plastic is a symptom of mass consumerism.
- Consumerism is using up resources at a massive rate.
- Plastic creates climate change emissions at every stage in its life cycle.
- Transporting, treating and disposing of plastic waste creates climate emissions.
- We need to stop depending on plastic.

In May 2019, the Center for International Environmental Law released a report – the first of its kind that deals with plastic and climate change – that suggested plastic, throughout its lifecycle, could contribute up to 13 per cent of the global carbon dioxide budget by 2050.

According to the report, the production and incineration of plastic waste in 2019 was due to produce more than 850 million metric tons (Mt) of greenhouse gases, which is equal to the emissions from 189 five-hundred-megawatt coal power plants working at full capacity. By 2050, if production continues as predicted (and planned) the amount will increase to the equivalent of 615 coal-fired power plants, producing 2.80 Gt of greenhouse gasses.

Plastic refining, the report says, is among the most greenhouse gas-intensive industries in the manufacturing sector – and the fastest growing. The manufacture of plastic uses a lot of energy and produces significant emissions through the chemical refining processes.

We've been so concerned with the oceans that we've forgotten, almost, the impact of making it, well before it's even got to the ocean as discarded waste.

The report from CIEL follows on from a different report, from the University of Hawaii at Manoa in 2018. Researchers there had discovered, unexpectedly, that the most common plastics, when exposed to sunlight, produce the greenhouse gases methane and ethylene. Polyethylene (LDPE), used in shopping bags, is the most produced and discarded synthetic polymer globally and was found to be the most prolific emitter of both gases.

The team also found that 'the emission rate of the gases from virgin pellets of LDPE increased during a 212-day experiment' and that 'LDPE debris found in the ocean also emitted greenhouse gases when exposed to sunlight.'

The study also raised concerns about microplastic disabling the ocean's ability to absorb carbon. At the surface, microscopic plants (phytoplankton) and animals (zooplankton) transfer carbon to the deep ocean: 'These plankton are being contaminated with microplastics and microfibres, which lab tests suggest reduce their ability to fix carbon through photosynthesis, perhaps also reducing their metabolic rates, reproductive success and survival rates.'

So if you ever needed a reason to choose natural fibres over nylon, carry a water bottle or refuse a straw, this is it.

And while it might seem to be too big or too difficult an issue to solve, it is one that we CAN do something about.

Time to Act:

Redirect your investments: if you have any kind of savings, investments or pensions, they may, ultimately, invest in fossil fuels and, by association, the plastics industry, so supporting them in their work. By divesting oneself of these kinds of industries you can help to pull the rug from underneath them, remove part of their funding and send a very clear message that you do not condone their practices.

REVISITING THE BEACH AT CLOVELLY

The bottle that was picked up on the beach in Clovelly will get recycled so it doesn't turn into thousands of pieces of microplastics. That's the good news. But the energy used to recycle it will add to our carbon deficit, as will the gases it emits during the rest of its time on earth. It will be turned into something else, of lesser purity and integrity that will need to be dealt with too, at some point. It will, eventually, get burned for energy, or put into landfill by future generations to deal with.

Unless we stop buying the bottles, consuming what's inside in minutes and then discarding them, the cycle will continue.

Plastic production is set to increase. The only way we can stop it entering the oceans is by turning off the tap.

We need to stop using so much plastic.

We need to consume less, reduce the demand for plastics and reduce our waste.

It's easy for us to be complicit in climate change and the plastics crisis, simply by not bothering to make the more difficult of the choices presented to us. The status quo – companies sending out their products encased in plastics, transporting it, creating plastic waste, investing in lucrative new markets in the developing world, funding recycling, us buying it because it's cheap and we like it, and carrying on as before – suits them down to the ground. It's a free pass to continue to make billions upon billions at our expense and at the expense of the planet.

And only you and I have the power – or the will – to do anything about it through our individual and collective actions.

Shall we? No more excuses.

CONCLUSION

Reading all this, and the rest of this book, you might be forgiven for thinking that the world's problems are insurmountable. But while you might not be able to replant the rainforests or refreeze the Arctic tundra, you can tackle your plastic consumption and reduce your waste. And that's the beauty of it. It's real and it's tangible and you can do something about it. Right here, right now.

There's no need to feel helpless either, because you'll be able to see how much of what you do matters. It's our problem and we have to fix it. And, in fixing it, we'll help to solve all kinds of other problems too.

Good luck and thank you for making a start.

Any more excuses? I hope not.

1. Recycle more and recycle responsibly; get to know how to recycle in your borough.
2. Eradicate as much single-use plastic as you can from your daily life.

3. Eat sustainably caught fish and less meat.
4. Buy local, fresh, seasonal, sustainable and packaging free, whenever you can.
5. Cook more and avoid pre-packaged meals.
6. Buy clothes only when you need them, and choose second hand or ethical whenever you can, in natural materials.
7. Get outside and enjoy nature more. It's free.
8. Spend your money on experiences, not things.
9. Resist the urge to buy the latest thing if you don't need it.
10. Buy, if you must buy, from businesses that are transparent about how things are made. But question their integrity first.
11. Think 'end of life' whenever you buy something. What will happen to it once you have finished with it?
12. Use your voice to make change happen.
13. Understand how much of what you do matters. It does.

SIMPLE PRODUCT SWAPS

I have included this list as a starting point for people who want to make a difference by making changes in their lives. It's about finding alternatives for the standard items that we are so used to using and can so easily continue to use without thinking.

While there are some great products out there to help you on your journey to plastic conscious living, I would like to add a vital caveat. As with a mobile phone, a china mug, a cotton bag made out of an old sheet, your favourite leather shoes or a fountain pen you've cherished for years, the greenest item you can use to ring someone, drink out of, pack loose vegetables in, wear out or write with is always going to be the thing you already have. The more you use it the smaller its footprint will be over time, and the better for the environment.

While it's fantastic that companies are making products to try to solve the plastic problem, some of the answers are already right in front of us. If it can be reused, repaired or reinvigorated then it's the best solution. If it can't be then you should **reconsider whether you really need it** before you buy it.

The last thing the planet, and you, need is more stuff.

IN THE BATHROOM

Shaving: in 2016 I ditched plastic razors for good, buying a safety razor with blades for about £10, a brush for £20 and a tin of shaving soap for about £5. For a total spend of about £35 I got a complete shaving kit that's lasted me until now. I use about a pack of blades a year (six), which is the only waste shaving creates for me now. They go in the recycling with the tins.

A subscription to Gillette's Fusion service would have cost around £210 for the same period and produced about 75 plastic razors with metal blades that cannot be recovered. It isn't the best a man can get. You don't need it.

Cleaning your teeth: plastic toothpaste tubes are hard to recycle and often won't be collected as part of your kerbside recycling. On top of that, around 3.6 billion toothbrushes are used each year globally and most of them won't be recycled. I find them often on beach cleans.

I have struggled with trying to make my teeth cleaning totally plastic-free. I have tried tablets, Holland and Barret's charcoal powder and Georganics coconut oil toothpaste. However, I have reverted to Euthymol, which comes in a metal tube and can be put into your metal recycling (although the lid is plastic) because I have had issues with sensitivity. I use a soft bamboo toothbrush from NOMAD Genie, which works for me.

Perfume: perfume bottles are notoriously hard to recycle because they are often made with composite materials. The Perfume Shop will now take your old glass bottle (and the lid and all the bits and pieces) and recycle it for you. And they will give you 10 per cent off your next purchase in store on the day (that's the cynical bit) www.theperfumeshop.com/About-Us.

Soaps and shampoos: liquid soaps that have dispenser tops are hard to recycle because they are composite. Solid soap bars do away with it entirely. If hygiene is an issue, try soap on a rope. Try www.theclovellysoapcompany.com.

For solid shampoo and conditioner try uk.lush.com. Their Naked range comes without plastic.

Also, www.beautykubes.co.uk make 'single serve' soaps, shampoos and conditioners that are perfect for travelling.

Cotton bud sticks: use paper stemmed versions.

Wipes and pads: there are plenty of people now making reusable and washable wipes from found materials and fabrics like www.facebook.com/travellingstitchesuk.

Female hygiene products: the average woman uses at least 11,000 throwaway sanitary products in a lifetime. The Hey Girls menstrual cup is recommended by Bude's Refill shop as they have a 'buy one give one' scheme to donate to UK women suffering period poverty, plus loads of great positive period education. They last ten years so are extremely cost effective and pose no risk of toxic shock syndrome.

IN THE KITCHEN

Inline filters: can be bought for your taps at home to filter impurities as well as plastic: www.pureh2o.co.uk/pure-h2o-co-purifiers-remove-plastic-fibres-found-in-uk-tap-water.

The Life Straw bottle claims to filter 99.999 per cent of plastic from water (www.lifestraw.com).

Chocolate and sweets: lindt still make their Excellence range with tin foil and paper as do many other brands. Travel sweets from Simpkins or Cavendish and Harvey still come in metal tins.

Sauces: go for glass alternatives and make your own, stored in glass jars.

Crisps: try crisps from: twofarmers.co.uk to avoid packets that can't be recycled.

Household cleaning: use cleaning soaps, soapnuts or refill containers from waste-free shops (www.soapnuts.co.uk).

Pan scourers: coconut husk scourers work well and can be composted at home. Buy them at www.safix.co.uk.

Cling film replacement: try beeswax wraps to replace cling film. They will grip to bowls and are antibacterial and can be washed. Make your own or buy them at www.beeswaxwraps.co.uk.

Washing bags: for catching microfibres, the Guppy Bag is the best product on the market at the moment. Put all your man-made fibres in it to wash. Buy at www.beachclean.shop/product/guppyfriend-washing-bag.

Plastic bag replacement: a lot of plastic bags can be recycled, which is great, but wouldn't it be better if we didn't have to? What if there were none? Use a cotton bag for life for your shopping and vegetable bags for your loose veg and create zero waste! Find out how to make one at morsbags.com.

Straws: get metal ones from www.beachclean.shop/product-category/eco-straws.

Cutlery: buy cutlery and all kinds of things at www.wearthlondon.com.

Reusable coffee cups: there have been a few worries about formaldehyde recently. None from www.rcup.co.uk. They are made from old coffee cups (40 per cent) and are recyclable.

Water bottles: a must-have for anyone, we choose www.jedz.co.uk as they donate £1 for every bottle sold to clean water projects and are insulated (cold drinks stay cold for 24 hours, hot drinks stay hot for 12 hours). Sigg bottles donate profits to the #2minutebeachclean (sigg.com).

IN THE GARDEN

Plastic pots: many thousands of plastic pots get bought and stashed in gardens all over the UK. But now there's a new idea that does away with them with a biodegradable pot that can be put into the ground with the plants. See more at posipot.co.uk.

So much plastic gets used in gardening, but it needn't. From plastic mulches to pots, crates, string and poles, it's unnecessary. Check out gardeningwithoutplastic.com for ideas.

DIY AND HOMEWARES

Carpets, curtains and soft furnishings: carpets made from man-made materials shed fibres that will, inevitably, get walked out of the house and down the drains. Aside from changing them for wool (which is often much more expensive) it may help to wear slippers inside and avoid walking outside shoes into the house. Vacuuming may help to suck up the fibres, as long as the hoover bag is emptied into the household waste and not on the compost heap.

It's the same with soft furnishings, rugs and curtains. Check the label.

Paint: we've long known that it's not great to wash paint in the sink, unless it's 'eco-friendly', even when it comes to water-based paints. Many acrylic paints contain tiny plastic particles that will go straight to the ocean when they go down the drain, just like fibres from clothes. Wash paint in fresh water and let the sludge settle, then soak it up with cat litter or sand before putting it in the household waste.

Following the EU Paint directive in 2004, paints are required to have labels defining the VOC content (harmful fumes from paint).

IDENTIFYING PLASTIC TYPES AND THEIR USAGE

Manufacturers use a rating system to help identify plastics for recycling – there are seven groups, each with a different set of characteristics and usages. Here are the symbols to look out for:

CODE AND SYMBOL	PLASTIC TYPE	TYPICALLY USED FOR	PROPERTIES
01 PET	Polyethylene terephthalate	Soft drinks bottles, food trays	RECYCLABLE Clear, tough, sinks
02 HDPE	High-density polyethylene	Yogurt containers, shopping bags, milk bottles, shampoo, detergent and chemical bottles	RECYCLABLE Floats in water
03 PVC	Polyvinyl chloride	Blister packs, pipes and hose, clear food packaging	RECYCLABLE (sometimes) Considered to be the most toxic of all plastics Not recommended for food use

CODE AND SYMBOL	PLASTIC TYPE	TYPICALLY USED FOR	PROPERTIES
04 LDPE	Low-density polyethylene	Rubbish bags, squeezable bottles, cling film	RECYCLABLE Floats in water
05 PP	Polypropylene	Bottle caps, straws, food tubs	RECYCLABLE Floats in water
06 PS 06 PS-E	Polystyrene, expanded polystyrene	Plastic cutlery, video cases, CD cases, cups, plates	NOT EASY TO RECYCLE Leaches styrene, a suspected carcinogen Not recommended for food use
07 PC, OTHER	Polycarbonate resins and composite materials	Components, computers, electronics	NOT EASY TO RECYCLE The 'catch all' for any plastics that can't be categorised by the other 6 Not recommended for food use. Risk of containing BPA.

Thank You to: Dolly, Nick, Jodie, Heather, Alan, Claire and Tracey; The 2 Minute Team; Tim, Laura and all at PFD; Laura, Liz, Jonathan, Abi, Siofra, Alice, Catherine and all at Penguin; Adam, Kate and Andy, The 2 Minute Trustees.

1 3 5 7 9 10 8 6 4 2

Published in 2020 by Ebury Press an imprint of Ebury Publishing,
20 Vauxhall Bridge Road,
London SW1V 2SA

Ebury Press is part of the Penguin Random House group of companies whose addresses can be found at global.penguinrandomhouse.com

First published by Ebury Press in 2020

www.penguin.co.uk

A CIP catalogue record for this book is available from the British Library

ISBN 978 1 529 10572 8

Colour origination by Altaimage, London
Printed and bound in Great Britain by Clays Ltd, Elcograf S.p.A.